152
Advances in Biochemical Engineering/Biotechnology

Aims and Scope

This book series reviews current trends in modern biotechnology and biochemical engineering. Its aim is to cover all aspects of these interdisciplinary disciplines, where knowledge, methods and expertise are required from chemistry, biochemistry, microbiology, molecular biology, chemical engineering and computer science.

Volumes are organized topically and provide a comprehensive discussion of developments in the field over the past 3–5 years. The series also discusses new discoveries and applications. Special volumes are dedicated to selected topics which focus on new biotechnological products and new processes for their synthesis and purification.

In general, volumes are edited by well-known guest editors. The series editor and publisher will, however, always be pleased to receive suggestions and supplementary information. Manuscripts are accepted in English.

In references, Advances in Biochemical Engineering/Biotechnology is abbreviated as *Adv. Biochem. Engin./Biotechnol.* and cited as a journal.

More information about this series at http://www.springer.com/series/10

Jie Bao · Qin Ye · Jian-Jiang Zhong
Editors

Bioreactor Engineering Research and Industrial Applications II

With contributions by

Michael W.W. Adams · Christoph Albermann · Jie Bao
Ju Chu · Julia Martin del Campo · Weiliang Hou
Zhimin Li · Jihan Lin · Joseph A. Rollin
Georg A. Sprenger · Mutsumi Takagi · Julia Tröndle
Guan Wang · Yonghong Wang · Michael Weiner
Dirk Weuster-Botz · Hui Wu · Jianye Xia · Qin Ye
Xinhao Ye · Y.-H. Percival Zhang · Jian Zhang
Siliang Zhang · Yingping Zhuang

Springer

Editors
Jie Bao
State Key Laboratory of Bioreactor Engineering, School of Biotechnology
East China University of Science and Technology
Shanghai
China

Qin Ye
State Key Laboratory of Bioreactor Engineering, School of Biotechnology
East China University of Science and Technology
Shanghai
China

Jian-Jiang Zhong
School of Life Sciences and Biotechnology
Shanghai Jiao Tong University
Shanghai
China

ISSN 0724-6145 ISSN 1616-8542 (electronic)
Advances in Biochemical Engineering/Biotechnology
ISBN 978-3-662-48346-6 ISBN 978-3-662-48347-3 (eBook)
DOI 10.1007/978-3-662-48347-3

Library of Congress Control Number: 2015956207

Springer Heidelberg New York Dordrecht London

Printed on acid-free paper

Springer-Verlag GmbH Berlin Heidelberg is part of Springer Science+Business Media
(www.springer.com)

Preface

Bioreactor is the core of industrial biotechnology practice. Traditional bioreactors include enzyme reactors and microbial fermentors, but recent research activities on bioreactors have significantly extended to various scales of them whether in a large natural ecology system or at a micro-scale level of nano sizes. Although bioreactor and bioreactor technology seem to be a traditional area of biochemical engineering and biotechnology, recently some new types of bioreactors have still been developed and applied into industrial practice. New methodology concepts on bioreactor research have also been proposed on the basis of cell biology and metabolic engineering. Therefore, in this context, this special volume is dedicated to bioreactor engineering to make the old new, and six invited chapters ranging from novel design of bioreactors to methodologies of bioreactor/bioprocess analysis and optimization are included.

As well known, gas-liquid contacting is an important unit operation in chemical and biochemical processes, and the gas utilization efficiency is generally low in conventional gas-liquid contactors, especially for sparingly soluble gases such as O_2, carbon monoxide (CO) and hydrogen (H_2). Ye et al. [1] described a novel self-inducing reactor for gas phase substrate utilization to break the barrier on gas-liquid transfer. The gas self-inducing impeller is able to recycle gas in the headspace of a reactor to the liquid bulk without utilization of additional equipment like a gas compressor, and its gas utilization efficiency is significantly enhanced. The authors introduced the principle, design, characteristics of self-inducing reactors as well as their application to biotechnology. Such as bioreactor may find its significant impact in CO and H_2 utilization by biotechnological approach.

H_2 is currently an important commodity chemical with its global annual market of about 45 million tons per year and a steady increase in demand over the last decades. It has the potential to become the future dominant transportation fuel. Technologies such as in vitro synthetic pathway biotransformation, by using over 10 enzymes to catalyze unnatural catabolic pathways, enable the H_2 storage in the form of carbohydrates [2]. Therefore, in vitro H_2 generation represents a clear opportunity for novel bioreactor and system design. This system is demonstrated with the conversion of glucose to H_2 at a high rate and the H_2 production from glucose 6-phosphate at a greatly increased reaction rate [2].

Healthcare and biomedical engineering is another important area of applied biotechnology. Traditional cultivation engineering of animal cells was developed from 1980s to industrialize the production of animal cell products such as interferon, interleukins, cytokines, tissue plasminogen activator, and therapeutic antibodies. The widespread use and industrialization of regenerative medicine are being highly expected because of the recent rapid progress in basic life sciences concerning human stem cells. Because the control of proliferation and differentiation of human cells and their three-dimensional culture into tissue are necessary for the realization of regenerative medicine in which cells are the final product, some new technologies of cell processing engineering, different from the conventional animal cell cultivation technology, should be developed. The cell quality such as cell heterogeneity should be noninvasively estimated before transplantation to patients, because cultured cells are usually heterogeneous and most protocols of regenerative medicine are autologous system. Takagi [3] reviewed the cell processing engineering including automatic cell processing and noninvasive cell quality estimation of adherent mammalian cells for regenerative medicine.

Lignocellulosic biomass utilization and biorefinery is another hot topic in recent years around the world, while pretreatment is a crucial step for overcoming the recalcitrance of lignocellulosic biomass in biorefinery processing. Pretreatment also affects both upstream and downstream operations beyond itself, and the efficiency of a biorefinery process is highly dependent on pretreatment technology used, such as pretreatment method, severity, inhibitors generated, wastewater discharge, residues released, and the reactors used. High solid loading in biomass pretreatment becomes a reasonable option and trend in the future industrial pretreatment operation. Zhang et al. [4] summarized the types, geometry, and design principle of pretreatment reactors at high solid loading of lignocellulose, which provided useful information for choosing and designing reactors of high-solids-loading biomass pretreatment.

In the last decades, targeted metabolic engineering of microbial cells has become one of the major tools in bioprocess design and optimization, for which detailed knowledge about relevant metabolic pathways and their regulation is required. Since in vitro experiments cannot display bioprocess conditions and behaviors properly, process-data about the cellular metabolic state has to be collected in vivo. Weiner et al. [5] highlighted several possibilities to gain information about the metabolic state of microorganisms, and summarized various methods to conduct perturbation experiments, which usually are the basis for detailed metabolic analyses and can be divided into dynamic and steady-state experiments.

In recent years, more and more efforts have been made for improving yields, titers, and productivities of aimed products through bioprocess engineering strategies. Among them, optimization and scale-up method for industrial processes are of great importance. It is indeed a permanent task and challenge for biochemical engineers to discover and gain more knowledge of cell kinetics and bioreactor fluid dynamics, and of the interaction between these two parts, in order to accelerate the technology transition from laboratory study to real industrial application. Xia et al. [6] highlighted the methodology for process optimization and bioreactor scale-up by integrating fluid dynamics with kinetics. A comprehensive discussion on advantages and challenges of the model-driven scale-up method was also given.

Finally, we hope this book is providing some new fundamental knowledge as well as practical experience to colleagues in the field of biochemical engineering and industrial biotechnology, which will be helpful to the technology advancement. Here, we would like to thank all the contributing authors for their excellent collaboration, and our special appreciation goes to the Managing Editor Prof. Dr. Thomas Scheper, all the referees, and the publisher and the book-series editorial staffs at Springer for their constructive suggestions, continuous support and kind help during the entire process from this special volume preparation to its final publication.

Shanghai
October 2015

Jie Bao
Qin Ye
Jian-Jiang Zhong

References

1. Ye Q, Li ZM, Wu H (2015) Principle and performance of gas self-inducing reactors and applications to biotechnology. ABE, doi: 10.1007/10_2014_329
2. Rollin JA, Ye XH, Campo JM, Adams MWW, Zhang YHP (2015) Novel hydrogen bioreactor and detection apparatus. ABE, doi: 10.1007/10_2014_274
3. Takagi M (2015) Cell processing engineering for regenerative medicine. ABE, doi: 10.1007/10_2014_282
4. Zhang J, Hou WL, Bao J (2015) Reactors for high solid loading pretreatment of lignocellulosic biomass. ABE, doi: 10.1007/10_2015_307
5. Weiner M, Tröndle J, Albermann C, Sprenger GA, Weuster-Botz D (2015) Perturbation experiments: approaches for metabolic pathway analysis in bioreactors. doi: 10.1007/10_2015_326
6. Xia JY, Wang G, Lin JH, Wang YH, Chu J, Zhuang YP, Zhang SL (2015) Advances and practices of bioprocess scale-up. ABE, doi: 10.1007/10_2014_293

Contents

Principle and Performance of Gas Self-inducing Reactors and Applications to Biotechnology 1
Qin Ye, Zhimin Li and Hui Wu

Novel Hydrogen Bioreactor and Detection Apparatus 35
Joseph A. Rollin, Xinhao Ye, Julia Martin del Campo, Michael W.W. Adams and Y.-H. Percival Zhang

Cell Processing Engineering for Regenerative Medicine 53
Mutsumi Takagi

Reactors for High Solid Loading Pretreatment of Lignocellulosic Biomass 75
Jian Zhang, Weiliang Hou and Jie Bao

Perturbation Experiments: Approaches for Metabolic Pathway Analysis in Bioreactors 91
Michael Weiner, Julia Tröndle, Christoph Albermann, Georg A. Sprenger and Dirk Weuster-Botz

Advances and Practices of Bioprocess Scale-up 137
Jianye Xia, Guan Wang, Jihan Lin, Yonghong Wang, Ju Chu, Yingping Zhuang and Siliang Zhang

Index 153

Adv Biochem Eng Biotechnol (2016) 152: 1–33
DOI 10.1007/10_2015_329

Published Online: 21 August 2015

Principle and Performance of Gas Self-inducing Reactors and Applications to Biotechnology

Qin Ye, Zhimin Li and Hui Wu

Abstract Gas–liquid contacting is an important unit operation in chemical and biochemical processes, but the gas utilization efficiency is low in conventional gas–liquid contactors especially for sparingly soluble gases. The gas self-inducing impeller is able to recycle gas in the headspace of a reactor to the liquid without utilization of additional equipment such as a gas compressor, and thus, the gas utilization efficiency is significantly enhanced. Gas induction is caused by the low pressure or deep vortex at a sufficiently high impeller speed, and the speed at which gas induction starts is termed the critical speed. The critical impeller speed, gas-induction flow rate, power consumption, and gas–liquid mass transfer are determined by the impeller design and operation conditions. When the reactor is operated in a dead-end mode, all the introduced gas can be completely used, and this feature is especially favorable to flammable and/or toxic gases. In this article, the principles, designs, characteristics of self-inducing reactors, and applications to biotechnology are described.

Keywords Critical speed · Gas-induction rate · Gas self-inducing reactor · Impeller · Power consumption · Volumetric mass transfer coefficient

Contents

1 Introduction 2
2 Principals of Gas Self-induction 3
3 Designs of Gas-inducing Impellers 4
  3.1 Some Self-inducing Impellers Disclosed in Patents 4
  3.2 Type 11 Impellers 5
  3.3 Type 12 Impeller 7

Q. Ye (✉) · Z. Li · H. Wu
State Key Laboratory of Bioreactor Engineering, East China University of Science and Technology, 130 Meilong Road, Shanghai 200237, China
e-mail: qye@ecust.edu.cn

3.4 Type 22 Impeller ... 8
3.5 Other Types of Gas-inducing Impellers ... 10
3.6 Stator ... 11
4 Performances of Self-inducing Reactors ... 11
4.1 Critical Rotation Speed ... 11
4.2 Induced Gas Flow Rate ... 15
4.3 Power Consumption ... 16
4.4 Gas–Liquid Mass Transfer Coefficient ... 18
4.5 Performance Prediction and Simulation ... 24
5 Applications in Biotechnology ... 26
6 Conclusion ... 28
References ... 29

1 Introduction

Gas–liquid contacting is an important unit operation in chemical and biochemical processes. The most popular operation mode of gas–liquid contacting is to continuously introduce a stream of compressed gas into a reactor below a rotating agitator, and the performances of such aerated and agitated reactors have been extensively studied [1, 2]. However, the gas utilization efficiency is generally low per gas passage through the liquid, especially when the solubility of gas substrate is low. For flammable or toxic gases, release of them into the environment is not allowed. To efficiently utilize such gas substrate, recycling the unreacted gas from the headspace of a reactor into the liquid is desirable, and a gas self-inducing impeller has such ability. Use of gas self-inducing impeller in a reactor does not require additional gas compressor or other device to recycle the gas substrate, and it can work efficiently with low power consumption and high mass transfer rate. Gas-inducing reactors have been used in froth flotation for a long time, as well as in chemical reactions such as hydrogenation, alkylation, and ozonization. Bioreactor equipped with a self-inducing agitator has been applied in bioprocesses, especially in vinegar production. In 1959, Hromatka and Ebner [3] reported repeated vinegar fermentation by using a self-inducing bioreactor. Matsumura et al. [4] proposed to recycle air in fermentation to satisfy the high oxygen demand in high cell-density culture.

The performance of gas-inducing reactors has been investigated extensively. Among the researchers, the team of J. B. Joshi has contributed greatly. Patwardhan and Joshi [5] have given a comprehensive review. The present article aims at the principles and performances of gas self-inducing reactors, and their potential application in biotechnology.

2 Principals of Gas Self-induction

A simple gas self-inducing impeller is composed of two or more hollow cylindrical blades with an orifice near the tip on each blade. The hollow impeller is connected to a hollow shaft with gas inlet holes above the liquid phase (Fig. 1a). When the impeller is in a static state, the liquid level inside the hollow shaft is the same as that in the reactor. When the impeller rotates, the liquid moves at a velocity relative to the impeller blade. According to Bernoulli's equation, this velocity results in a lower pressure at the orifice of the impeller blade, and the liquid level in the shaft lowers to some extent. When the impeller rotates at a sufficiently high speed, the liquid level in the hollow shaft is lowered to the level of blade. This speed is termed as the critical speed. At the critical speed, no gas induction occurs. If the impeller is rotated at a further higher speed, the pressure inside the orifice on the impeller blade is low enough to overcome the liquid static pressure and the resistances to the gas flow, and the gas in the headspace starts to be sucked and distributed into the liquid through the orifices on the cylindrical blades. A further increase in the impeller speed increases the gas-inducing rate.

The second type of self-inducing impeller is shown in Fig. 1b. The hollow impeller is composed of two disks, between which several vertical blades exist. This impeller is connected to a hollow shaft with gas inlet above the liquid surface.

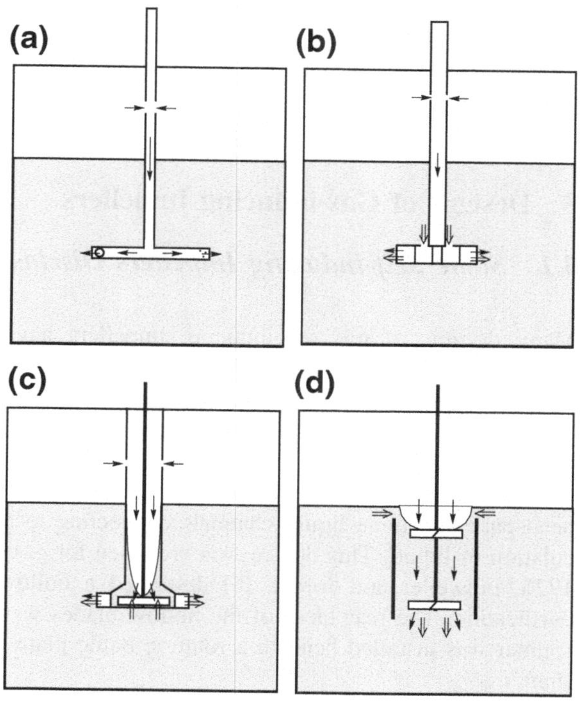

Fig. 1 Schematic diagram of self-inducing reactors. → gas flow; ⇒ liquid flow

There exists a gap between the shaft and the upper disk of the impeller. When the impeller rotates at a high speed, the low pressure at the impeller tips causes gas in the headspace to enter the hollow shaft. At the same time, some liquid enters the impeller through the gap, and both gas and liquid leave the impeller.

The third type of gas-induction impeller (Fig. 1c) is surrounded by a stator to disperse the induced gas to the liquid. The stator is at the bottom of a standpipe, whose upper end is open to the gas phase of the reactor. When the impeller is not rotating, the liquid level in the standpipe is the same as that in the reactor. As the impeller rotates, a vortex is formed in the standpipe, and the depth of vortex increases with the increase in the rotating speed of impeller. When the rotating speed is high enough, the vortex reaches the impeller and gas induction begins.

The fourth type of gas-induction impeller system is composed of an impeller near the liquid surface, and the impeller can be installed within a draft tube (Fig. 1d). The rotating agitator produces a vortex, and at a high enough rotating speed, the vortex reaches the upper impeller. The induced gas flows inside the draft tube and then is distributed to the liquid outside the draft tube. Multiple impellers can be fitted on the shaft, and the induced gas can be distributed in a reactor with a high aspect ratio.

Patwardhan and Joshi [5] have classified the gas-inducing impellers into three types based on the presence of gas and liquid phases in the impeller zone (Fig. 1): A, gas flow alone at both the inlet and outlet (type 11); B, gas only at the inlet but both gas and liquid at the outlet of impeller (type 12); and C, both gas and liquid at the inlet and outlet (type 22), of the impeller zone. For both the type 11 and type 12 impellers, the induced gas alone enters the impeller zone through a rotating, hollow shaft, and thus, these impellers are different from the type 22 impellers and the impeller shown in Fig. 1d.

3 Designs of Gas-inducing Impellers

3.1 Some Self-inducing Impellers Disclosed in Patents

Many designs of gas self-inducing impellers have been disclosed in patents. In 1877, Long and Castle [6] disclosed a hollow agitator for distribution of mercury-containing vapor in water to condense mercury. The impeller was composed of two hollow blades with trapezoid cross-section, and the blade tips were open to the liquid. Then, in 1918, Ruth [7] mentioned a rotating head with three channels connecting to a central upper gas inlet tube to introduce gas from the headspace and three liquid channels connecting to a central bottom inlet for circulation of liquid. This device was proposed for gas washing and humidifying. In 1926, Fulweiler and Jordan [8] described a hollow four-blade spumer for gas purification. The rear faces of the hollow blades were open to the liquid, and the spumer was installed beneath a rotating baffle plate mounted on the same hollow shaft.

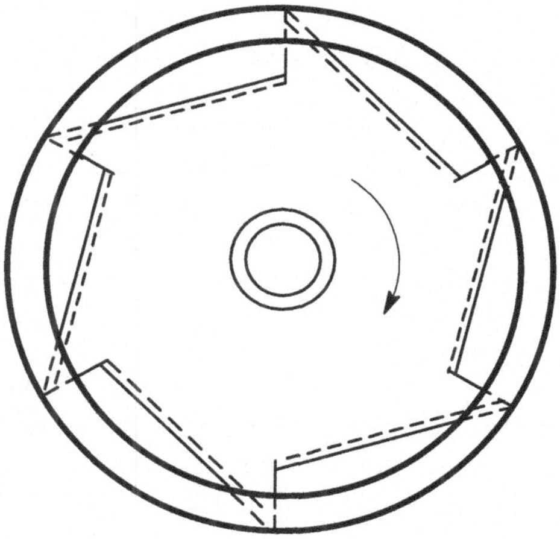

Fig. 2 Hollow ratchet-type impeller

In 1955, Enekel and Maurer disclosed a self-aspirating ratchet-type impeller wheel to simulate gas aspiration when liquid passes through a continuously tapering and then suddenly widen tube similar to a Venturi tube [9]. The ratchet wheel was a six-vaned impeller with openings for egress of air facing away from the direction of rotation (Fig. 2). A stator (not shown in Fig. 2), which was placed around the rotor, contained an upper ring and a bottom ring between which were 8 guiding plates to guide the flow of the gas–liquid mixture. When the ratchet wheel rotated at a high speed, air was sucked from the headspace through a hollow driving shaft and dispersed into liquid through the openings of the ratchet impeller to obtain efficient gas dispersion. This design was further improved by Ebner and Enenkel [10] who proposed to use a short hollow driving shaft connecting to a stationary gas pipe to introduce air from outside of the reactor. Such a design allowed the rotor to rotate at very high speeds (more than 1000 rpm) driven by a motor beneath the reactor. The rotor was covered with an upper and a bottom ring to guide the liquid flow inside the spaces between the rear opening and front vane. The rotor was surrounded by a stator composed of 12–36 guiding plates between two rings with large widths. The gap between the rotor and the stator was small, only 0.5–3 mm. The gas-induction rate could reach 200 m^3/h (0.65 vvm) in a commercial bioreactor with a diameter of 4 m and a liquid height 4.1 m, and the specific power consumption was only 0.085 kw/(m^3 air).

3.2 Type 11 Impellers

This impeller system is composed of a hollow stirrer connecting to a hollow shaft with gas inlets above the liquid surface. The simplest self-inducing impeller of this type is shown in Fig. 1a and has been extensively studied [11, 12]. The gas-induction rate is dependent upon the number of impeller blade. Zlokarnik and

Judat [13] investigated a hollow impeller containing 4-pipe blades with tips of 45° oblique cuts, and Heim et al. [14] studied impellers of four- and six-pipe blades with 45° cuts. It was found that use of more pipe blades improved gas induction. At the same rotation speed, the highest gas-induction rate was obtained for the 6-pipe impeller, followed by the 4-pipe impeller, with the lowest for the 2-pipe impeller [12]. The resistance encountered by gas flowing through the hollow shaft and impeller affects gas induction. To increase the gas-induction rate, hollow blades with multiple orifices or with larger orifice areas were used [12, 15]. Increasing the number of gas outlet resulted in increased gas-induction rate at the same rotation speed and higher mass transfer coefficient $k_L a$ at the same power consumption per unit liquid volume. As alternatives, hollow concave impeller [16, 17], flattened cylinder impeller [18], hollow pitched-blade impeller with multiple orifices [19], and ratchet-type disk impeller [14] have also been studied.

In addition to these impellers, other hollow agitators have also been proposed. Zlokarnik [20] investigated a series of hollow agitators with the shapes of triangle, square, pentagon (Fig. 3), and hexagon, near the vertexes of which gas egress holes were drilled on the side against the moving direction. Furthermore, round hollow wheels were also used. Recently, Achouri et al. [21] used a hollow wheel with six gas egress holes but without any blades for gas induction. Since no blades existed, the power consumption was low.

The submergence of the gas-inducing impeller is an essential factor that affects gas induction significantly. To obtain a high gas-induction rate, the submergence cannot be too great. However, this may result in poor mixing in the reactor. Increasing the extent of liquid circulation is beneficial to the mass transfer through gas–liquid interface, and use of multiple impellers can improve the dispersion of gas–liquid mixture and thus improves the performance of gas induction. For example, Mohammed et al. [22] used a shrouded turbine with curveted blades for

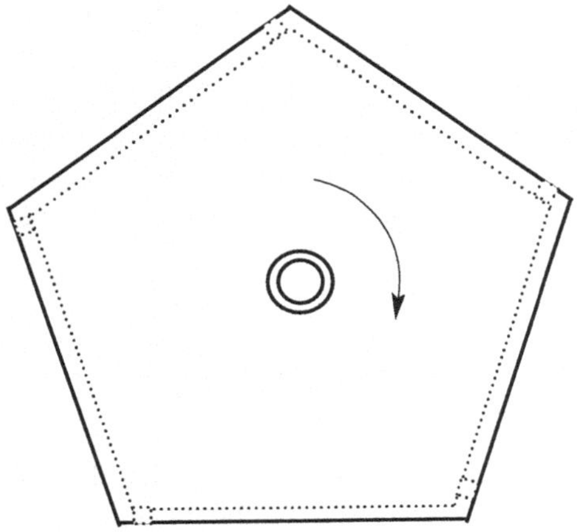

Fig. 3 Hollow pentagon impeller

gas induction, and a lower disk turbine for mixing the gas–liquid mixture. Kasundra et al. [23] studied the effects of combinations of different types of gas-inducing impellers and liquid circulation impellers. Their upper gas-inducing impellers included downward flow hollow pitched-blade turbine (PBTD, 45 or 60°), modified double-disk (MDD) turbine, six-pipe impeller, and modified four-pipe impeller; the bottom impellers included six-blade hydrofoil, PBTD, and upward flow pitched-blade turbine (PBTU, 45°). They found that the MDD turbine was most energy efficient, and the combinations of MDD plus hydrofoil and hollow PBTD plus hydrofoil were recommended. Ju et al. [24] also studied two-impeller gas-inducing reactor, in which the upper impeller for gas induction was a standard retreat blade radial turbine with six orifices on the shaft for gas egress. The gas outlets were extended by fixing three or six pipes with the lengths larger than the radius of the turbine to lower the critical impeller speed. Recently, Gomadurai et al. [25] studied a gas-inducing impeller with the similar design of Ju, but the gas-inducing pipe was bent to L-shape.

Joshi's team further studied pitched turbine downflow (PTD) impellers in a draft tube. Mundale and Joshi [26] found that impellers with 6 blades resulted in best air-inducing effects, followed by blade number of 7, 5, 4, 8, 3, and 12, and an even number blade impeller was better than one with odd number blades. Holes on the blades deteriorated the air-inducing performance unless the holes were located at the tip. The blade curvatures practically did not bring about any influence.

Wu et al. [27] used a hollow self-inducing bioreactor for recycling CO_2 in succinic acid fermentation (Fig. 4). They did not use a hollow shaft but a concentric gas introduction tube fixing to the shaft. The diameter of the gas introduction tube was much larger than that of the shaft to reduce the gas flow resistance. The upper part of the tube was open to the headspace, but the bottom was closed. Six hollow blades with tips of 45° oblique cuts were connected to the gas tube near the bottom to allow the flow of gas. Two rings were placed on the hollow blades to guide the liquid flow. At a speed of 300 rpm, gas induction could be observed.

3.3 Type 12 Impeller

This type of impeller also contains a hollow shaft and a hollow impeller, but differs from the type 11 impeller in the presence of a liquid phase in the impeller zone. Joshi drilled four holes on the hub of a 2-pipe impeller to allow liquid entering the pipe impeller [28]. This modification increased the gas-induction rate by 30–45 % compared with that without liquid inlets. The increase in gas-induction rate was attributed to increased apparent density due to entered liquid phase. Recently, Wang et al. [29] studied a 6-pipe impeller with a liquid inlet hole on each pipe near the shaft and found that the liquid flow inside the pipe impeller resulted in decreased critical impeller speed and power consumption and increased gas holdup.

Fig. 4 Hollow 6-blade impeller with a large gas introduction tube

3.4 *Type 22 Impeller*

The schematic diagram of the reactor equipped with this type of impeller system is shown in Fig. 1c. This impeller system is used in many flotation cells such as Wemco, Wedag, Denver, Agitair, Booth, and Sala [5]. Sawant and Joshi investigated the critical impeller speed of Wemco and square Denver gas–liquid contactors [30], and the characteristics of mass transfer and hydrodynamic properties of the Denver-type flotation cells [31]. It was found that the performances of Denver contactor were similar to conventional aerated and agitated contactors, but Wemco contactor was better than Denver contactor. Zundelevich [32] studied the performances of a turbo aerator equipped with a hollow turbine which was composed of a central disk with radial blades mounted between two rings. This type of aerator gave much higher gas-induction rate than Denver flotation cell based on the power consumption per unit liquid volume [33]. Mundale and Joshi [34] examined the effects of different types of impellers on gas induction and concluded that the downflow pitched 6-blade turbine (PBTD) was the most energy efficient, and the pipe, modified pipe, and flattened pipe impellers were also better than Denver cell [33, 35].

Enhanced mixing of the gas–liquid mixture is beneficial to improvement of gas–liquid contacting. Therefore, researchers have investigated reactors equipped with multiple impellers. The two-impeller system includes an upper impeller for gas induction and a bottom impeller for dispersion of gas–liquid mixture. Saravanan and Joshi [36] studied the characteristics of reactors equipped with a gas-inducing PBTD, in combination with different impellers including PBTD and PBTU, disk turbine, downflow propeller, and upflow propeller. Based on power consumption, the bottom upflow propeller gave the best gas induction, followed by PBTU, while disk turbine was the poorest. When a bottom PBTU impeller was used, the induced bubbles were more perfectly dispersed in the bottom region of reactor with increasing agitation speed [37]. Power consumption was related to the impeller design, submergence, and clearance to the bottom of reactor [38]. Introduction of an air stream through a sparger resulted in an increase in critical speed and a decrease in gas-induction rate [39].

Mundale and Joshi [34] found a hysteresis phenomenon: The gas-induction rate suddenly decreased when the impeller speed was increased to a high speed, which is called the first transition speed. At this state, further increase in impeller speed could not greatly increase the gas-induction rate. When the impeller speed was decreased to one called second transition speed, the gas-induction rate recovered to the original state. Patil and Joshi [40] further studied factors affecting the stability of gas induction and attributed the abrupt decrease in gas-induction rate at the first transition speed to: (i) With the increase in impeller speed, the liquid level became lower inside the impeller zone. At the transition speed, the entire impeller was exposed to the gas phase, resulting in a great drop of the pumping capacity of the impeller and sudden reduction in gas-induction rate. (ii) The low liquid level resulted in reversal liquid flow in the upper part of stator and drowned the impeller to reduce the gas-induction efficiency.

Poncin et al. [41] studied a type 22 impeller, which was a hollow turbine consisted of eight blades enclosed by two disks, and the upper disk was a ring plate (Fig. 5). The open annular area between the upper ring and the shaft allowed liquid to enter the impeller zone. At a high enough rotation speed, the pressure at the joint of impeller and hollow shaft is lower than the static liquid pressure, and gas in the headspace of vessel is sucked to enter the spaces between blades. At the same time, liquid enters the impeller zone through the annular area of the upper disc.

Sardeing et al. [42, 43] studied a commercial available agitator system, Turboxal, which contained three impellers: a shrouded gas-inducing impeller,

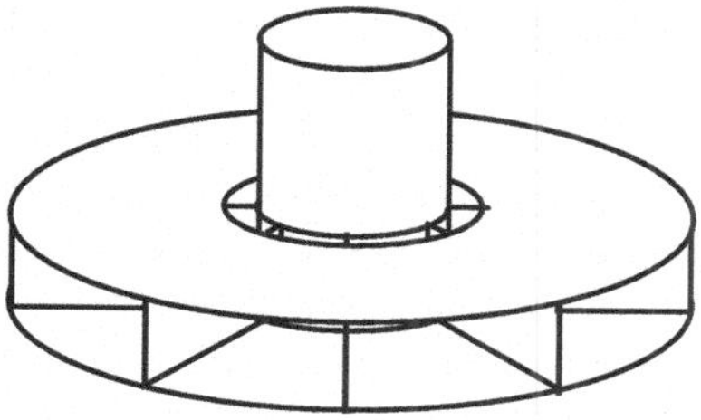

Fig. 5 Hollow disk impeller with liquid inlet

an upflow propeller, and a downflow propeller. The gas-inducing impeller was similar to that used by Poncin et al. [41] and was connected to a hollow shaft and covered by a shell for directing the flow of gas–liquid mixture. This gas-inducing system was situated very near the liquid surface, and the diameter of the impeller was small as compared with the vessel. It has a high flooding rate when an oxygen stream is introduced and is suitable to wastewater biotreatment [44]. Sardeing et al. [45] further investigated the commercial available gas-inducing impeller and made a series of modifications, which were made on the bottom disk by replacing with a ring leaving the center open to liquid (the crown turbine, TC), on the upper ring by changing the width of the annular space (mixed turbines TMW and TMP), and reducing or even eliminating the bottom disk (TMD). Computation indicated that the TMW5 and TMD50 gave better pumping efficiency than the original design.

3.5 Other Types of Gas-inducing Impellers

An alternative design is gas entrainment by an impeller just submerged below the liquid surface (Fig. 1d). At a sufficiently high impeller speed, the formed vortex reaches the impeller and gas above the liquid surface is entrained into the liquid. This type of gas induction is sometimes called as surface aeration when the gas is air, and Patwardhay and Joshi [46] have made a detailed review on this type of reactors. Matsumura et al. [47] proposed a fermentor for high cell-density culture of microbial cells with pure oxygen which can be used efficiently. The fermentor was equipped with an upper propeller near the liquid surface to entrain the oxygen gas in the headspace and a bottom agitator of pitched-blade impeller to disperse the entrained gas. The rotation of both impellers was controlled separately. Their research indicated that the gas-entraining propeller affected the oxygen consumption rate especially at low agitation rate of the bottom impeller. They also studied the critical speed of the bottom impeller and correlations of gas-entraining capacity and power consumption [4]. However, no results of real fermentation experiments have been reported. Hsu and Chang [48] found that reduction of the baffle width deepened the vortex, and its effect was greater than the submergence of the upper impeller. This type of reactor was modified by placing a draft tube around the impellers, and the baffles were moved to the inner wall of draft tube [49, 50]. The mass transfer performances were studied by using ozone to oxidize the dye Reactive Blue 19 [49], an ozone sensor [50], or a DO sensor [51].

When multiple impellers are fitted on the same shaft, a reactor with a large height/diameter ratio can be constructed. Scargiali et al. [52] studied a reactor with an aspect ratio of 5.18 (internal diameter of 280 mm and height of 1450 mm). The reactor contained a draft tube, inside which were four vertical baffles. Five pitched-blade impellers were mounted on a shaft, and at the bottom of reactor, a six-blade radial turbine was installed to recycle the gas–liquid mixture. They estimated power consumption by measuring the temperature increase rate and determined the mass transfer coefficient k_La with the dynamic method in four different modes. Among these modes, the pure oxygen–vacuum-degassed liquid mode gave the highest $k_{L}a$ values compared with the modes of pure oxygen–air saturated liquid, air–oxygen saturated liquid, and air–nitrogen saturated liquid. The presence of solid particles resulted in increased critical impeller speed and power consumption, as well as decreased $k_{L}a$ [53].

Researchers have also constructed reactors with a helical ribbon impeller to entrain gas. Kamen et al. [54] investigated a bioreactor equipped with a helical ribbon impeller and vertical surface baffles for culture of shear-sensitive insect cells. The diameter of the helical impeller was a little smaller than the tank diameter. At an impeller speed of 75 rpm, gas bubbles started to be observed in the liquid phase. Conway et al. [55] constructed a reactor equipped with a helical screw impeller inside a central draft tube for gas–liquid–solid dispersion. Underneath the helical impeller, there was a radially bladed impeller. In addition, cruciform baffles were placed at the upper and bottom ends of the draft tube, and baffles were also placed on the wall of reactor. This type reactor was considered suitable for solid suspension.

3.6 Stator

The stator plays an important role in dispersing the gas–liquid mixture in a gas self-inducing reactor. In the design of Enenkel and Maurer [9], the ratchet rotor was surrounded by a stator. The upper ring of stator covered part of the rotor. Ebner and Enenkel [10] further suggested a stator with large ring diameter, 12–36 stator blades, and small stator blade angle with radius.

For the initial design of type 22 impeller, a stator without a lower ring is mounted at the bottom of the standpipe to guide the flow of the gas–liquid mixture. Another design is to place the liquid entrances on the shoulder of standpipe. However, in Zundelevich's design of turbo aerator [32], no liquid inlet was placed on the standpipe and the stator blades were mounted between two rings. Patil and Joshi [56] studied the effects of various stator designs such as number of blade (0–24), shape of blade (vane or rod), width of stator ring, height of blade, and radial gap between the rotor and the stator in reactors with geometrical similarity.

4 Performances of Self-inducing Reactors

4.1 Critical Rotation Speed

All the gas-inducing reactors operate at an agitation speed higher than the critical speed, N_{C}, whatever the type is. The critical impeller speed is dependent upon the impeller design and the depth of impeller submergence. Studies on prediction of the critical speed have been carried out for the simplest gas-inducing impeller consisting of a hollow shaft and a two-pipe (round, flattened, or concave) impeller with openings at the blade tips [11, 12, 16, 18, 57]. Gas induction happens due to the pressure reduction caused by the relative speed between the orifice and liquid. Evans et al. [57] derived the equation for calculating the critical impeller speed based on Bernoulli's equation:

$$\mathrm{d}\left(\frac{v^2}{2}\right) + \mathrm{d}\frac{p}{\rho_{\mathrm{L}}} = 0 \tag{1}$$

where v is the orifice velocity, p is the combined hydrostatic and pressure head, and ρ_{L} is the liquid density. The liquid velocity relative to the impeller orifice u can be related to the impeller rotating speed N by introduction of a slip factor, K:

$$u = 2\pi Nr(1 - K) \tag{2}$$

where r is the distance of orifice from the center of shaft. The value of K has been determined by many researchers, e.g., 0.9–1.3 for pipe and flattened cylindrical impellers [11] and 1.15 for a turbo aerator [32]. A pressure coefficient C_p is introduced to define the pressure difference between the stagnation pressure, p_{S}, and the local pressure at the orifice, p_{l}:

$$C_p = \frac{p_{\mathrm{S}} - p_{\mathrm{l}}}{\rho_{\mathrm{L}} u^2/2} \tag{3}$$

The value of C_p is dependent on the position of orifice and the design of impeller, and $C_p = 4\sin^2\theta$, where θ is the angle from the rear stagnation point [57]. From Eqs. (1) to (3), the following equation can be obtained [11]:

$$p_{\mathrm{l}} = p_{\mathrm{O}} + \rho_{\mathrm{L}} gh + \frac{1}{2}\rho_{\mathrm{L}}(1 - C_p)[2\pi Nr(1 - K)]^2 \tag{4}$$

where p_{O} is the gas pressure in the headspace and h is the submergence of impeller. At the critical impeller speed, p_{l} equals to the headspace pressure and from Eq. (4) the critical impeller speed is as follows:

$$N_{\mathrm{C}} = \sqrt{\frac{gh}{2(C_p - 1)[\pi r(1 - K)]^2}} \tag{5}$$

To start gas induction, the gas should overcome the radial pressure drop in the horizontal pipe blade $[2\rho_{\mathrm{L}}(\pi Nr)^2]$, even when the meniscus in a U tube connecting to the lower end of the hollow shaft lowered to the level of the orifice [57].

For the impeller systems with a standpipe (namely type 22), gas induction is started when the vortex in the standpipe reaches the impeller. White and de Villiers studied gas induction in a laboratory flotation machine consisted of a 12-blade rotor and a stator with 16 slots at the height of the impeller blades [58]. At the critical impeller speed, they showed that the pressure drop across the vortex

$$\rho_L gh = k\rho_L N_C^2 D^2 \tag{6}$$

where k is a constant, and D is the impeller diameter. Therefore,

$$Fr_C = \frac{N_C^2 D^2}{gh} = \frac{1}{k} \tag{7}$$

where Fr_C is the (modified) Froude number at the critical impeller speed. In his study on the gas-inducing rate of the turbo aerator, Zundelevich [32] analyzed gas induction by analogy of a jet injector and proposed a proportional relationship between the gas flow rate and the liquid flow rate. For the impeller design he used, the proportional constant ψ (the injection coefficient) was derived to be

$$\psi = 3\sqrt{1 - 3\frac{2/Fr}{k_1\pi^2 + 2/Fr}} - 1 \tag{8}$$

where k_1 is a constant. He pointed out that Eq. (8) could not be used for reliable prediction of the injection coefficient due to lack of distinctly pronounced injector elements in the self-inducing aerator design, but its significance was to show that ψ depends upon Fr alone. Although in this study he did not analyze the critical impeller speed of the turbo aerator, the critical speed can be deduced from Eq. (8). Since at the critical speed the gas-induction rate is zero, but the liquid flow rate, which is proportional to ND^3, is not zero, the injection coefficient ψ must be zero. Thus, by setting ψ to zero, the following equation can be obtained:

$$Fr_C = \frac{19}{4K_1\pi^2} \tag{9}$$

Equation (9) cannot be used to predict the critical speed of the turbo aerator reliably, neither, but it indicates that the Froude number at the critical impeller speed is a constant like that pointed out by Eq. (7).

Sawant and Joshi examined the effects of physical properties of liquid on the critical impeller speed of square Denver flotation cell and round Wemco cell [30]. It was found that the density and surface tension did not affect N_C, but the liquid viscosity affected N_C:

$$\frac{N_C^2 D^2}{gh}\left(\frac{\mu_W}{\mu}\right)^{0.11} = \frac{1}{k} \tag{10}$$

where μ_W and μ are the viscosities of water and the liquid used, respectively. For different types of self-inducing reactors, the constant $1/k$ was calculated according to published literatures and the value was estimated to be 0.21 ± 0.04, indicating $1/k$ is practically independent on the impeller design and diameter. Saravanan et al. [35] analyzed the formation of paraboloid vortex in the standpipe and derived the equation:

$$N_C = \frac{1}{2\pi R}\left(\frac{2gf_{sp}(h - a_C)}{\phi_C(1 - I_C^2/R^2)}\right)^{\frac{1}{2}} \tag{11}$$

where R is the radius of impeller and f_{sp}, a_C, ϕ_C, and I_C are lumped constants. The values of these parameters were obtained from 107 sets of experimental data, and the equation could predict the critical speed within an error of 8 %. For a reactor equipped with a gas-inducing impeller coupled with a second impeller for gas–liquid mixture dispersion, they found that a slightly simplified form of Eq. (11) could well predict N_C [36]:

$$N_C = \frac{1}{2\pi R}\sqrt{\frac{2gh}{\phi}} \tag{12}$$

In a reactor equipped with a standpipe and multiple impellers, when a stream of gas is sparged, the critical speed increases. Patwardhan and Joshi [39] derived an empirical equation to correlate the increase in critical speed, ΔN_C, with gas velocity and reactor dimension:

$$\frac{\Delta N_C^2 \pi^2 D^2 \phi}{2gh} = \alpha_1 (C_I/D)^{\alpha_2} v_G^{\alpha_3} (h/T)^{\alpha_4} \tag{13}$$

where α_1, α_2, α_3, and α_4 are constants, C_I is the clearance between two impellers, v_G the superficial gas velocity, and T the tank diameter.

For liquid containing solid particles such as catalyst or microbial pellet, it is important to suspend the particles to obtain better gas–liquid–solid contacting. Since the submergence of the gas-inducing impeller is usually limited to obtain a high gas-inducing rate, the clearance to the bottom of reactor is large, resulting in poor suspending ability. Use of the second impeller is favorable to gas dispersion and solid particle suspension. Saravanan et al. [59] investigated solid suspension in reactors equipped with a gas-inducing PBTD inside a standpipe, combined with a bottom impeller out of five different types (PBTD, PBTU, disk turbine, upflow propeller, and straight blade turbine). The critical solid suspension speed (N_{CS}) was defined as the impeller speed at which solid particles were completely suspended and no particles stayed on the tank bottom for more than 1–2 s. The bottom impeller of PBTD gave the lowest N_{CS}, followed by disk turbine, straight blade turbine, and PBTU, while upflow propeller showed the highest N_{CS}. For the dual-PBTD combination, the critical suspension speed could be calculated by:

$$N_{CS} = \alpha_1 \sqrt{\frac{\rho_S - \rho_L}{\rho_L}} A^{-0.83} \left(\frac{W}{D}\right)^{-0.314} d_P^{\alpha_2} X^{\alpha_3} \left(\frac{C_3}{D}\right)^{\alpha_4} D^{\alpha_5} \left(\frac{C_1}{T}\right)^{\alpha_6} T^{\alpha_7} \quad (14)$$

where ρ_S and ρ_L are solid and liquid densities; C_1 and C_3 are the clearance of the bottom impeller to bottom and that between the impellers, respectively, A is the blade angle, W is the blade width, d_P is the particle diameter, and X is the solid loading. The parameters α_1 through α_7 were estimated from 166 sets of experimental data, and the error of the predicted N_{CS} was within 12 %.

4.2 Induced Gas Flow Rate

The gas-induction rate is affected by the submergence and rotation speed of the gas-inducing impeller and is also determined by the impeller design. Many studies have been carried out to calculate the induced gas flow rate. Empirical models were obtained by correlating the induced gas flow rate, Q, with operating variables. The simplest form is as follows:

$$Q \propto N^a D^b \quad (15)$$

where a and b are empirical constants. For example, their values were 1.5 and 3.0 determined by Sawant et al. [31] for Denver-type flotation cells. However, usage of empirical equations should be careful, while mechanistic models are more favorable. For the hollow pipe self-inducing impeller, Martin [18] derived an equation to predict the induced gas flow rate based on the equation of fluid flow rate through an orifice:

$$Q = C_O A_O K_1 \sqrt{2gh_s(\rho_L/\rho_G)} - K_2 \quad (16)$$

where C_O is the orifice coefficient, A_O the orifice area, ρ_G the gas density, and h_s the liquid head outside the orifice when no gas is allowed to flow, and K_1 and K_2 are constants. Similarly, Evans et al. [57]. derived an equation considering the head loss due to the resistances in the hollow pipes and through the orifices:

$$Q = C_O A_O \sqrt{\left\{C_P[2\pi r N(1-K)]^2 - 2gh\right\} \rho_L/\rho_G} \quad (17)$$

For pipe impeller with multiple gas egress orifices, Forrester and Rielly [15] derived the gas flow rate through the ith orifice, Q_{Oi}:

$$Q_{Oi} = \frac{4\pi u_i \sqrt{C_{pi}}(r_{bi}^3 - r_{oi}^3)}{3(r_{bi} + r_{oi})} \quad (18)$$

where u_i, C_{pi}, r_{bi}, and r_{oi} are the relative liquid velocity and C_p at, detached bubble radius and radius of, the ith orifice, respectively. This equation could well predict the gas-induction rate for one-orifice impeller, but over-predicted the gas rate with more orifices.

For the self-inducing impeller with a standpipe, it is hard to derive a mechanistic model. Therefore, empirical models are used. Raidoo et al. [33] obtained the following equation which predicted the gas-induction rate within an error of 10 %:

$$Q^2 = c(\Delta p \cdot D^2)^{1.184} \tag{19}$$

where Δp is the pressure driving force and c is a constant. Saravanan et al. [35, 36] derived a model to predict the gas-induction rate for a pitched-blade impeller based on the vortex in the standpipe. The total gas-induction rate was assumed to be composed of two items: bubble entrapment (Q_{E}) and carriage (Q_{C}), and the equations for Q_{E} and Q_{C} are as follows:

$$Q_{\mathrm{E}} = \lambda NR^2\left[1 - \frac{2gh}{\phi v^2}\right] \tag{20}$$

$$Q_{\mathrm{C}} = \alpha NR^3\left[1 - c\left(\frac{2gh}{\phi v^2}\right)\right]^{\frac{3}{2}} \tag{21}$$

where λ is gas-induction modulus, α is gas pumping constant, ϕ is vortexing constant, c is a constant, and v is impeller tip velocity. Thus, the total gas-induction rate Q could be calculated by:

$$Q = \lambda NR^2\left[1 - \frac{2gh}{\phi v^2}\right] + \alpha NR^3\left[1 - C\left(\frac{2gh}{\phi v^2}\right)^{\frac{3}{2}}\right] \tag{22}$$

and the values of λ, α, ϕ, and C were estimated for single pitched-blade impeller and different impeller combinations of two impellers [36]. Patil et al. [60] also investigated gas induction using the same model and a laser Doppler anemometer to understand the local liquid flow rate in a contactor with diameter of 0.5 m.

4.3 *Power Consumption*

Power consumption is a variable critical to reactor design. In an agitated reactor, power consumption for agitation can be expressed as follows:

$$P = N_P \rho N^3 D^5 \tag{23}$$

where P is power consumption, ρ is liquid density, and N_P is power number. N_P is dependent upon the impeller design and Reynolds number and is constant for turbulent flow. For example, the N_P value of a Rushton disk impeller in turbulent flow is 5. When a stream of gas is introduced, the power consumption decreased due to lowered liquid density and formation of cavities on the rare blade face of impeller. The reduction of power consumption is dependent on the reactor design, gas flow rate, and the properties of liquid. For a self-inducing reactor, the gas-induction rate is a function of impeller speed; thus, the impeller speed greatly affects power consumption. Many researchers have investigated the characteristics of power consumption in self-inducing reactors and derived empirical equations to correlate power required with operation variables and liquid properties. For example, Martin [18] indicated that the N_P for the cylinder impeller in a tank with three baffles was approximately 1.0. Joshi and Sharma [12] correlated the power consumption for a two-pipe cylindrical impeller as:

$$P = \left(1.0 - 17\frac{Q}{ND^3}\right) N_P \rho N^3 D^5 \tag{24}$$

with a standard deviation of 7 %, or as:

$$P = c\left(\frac{P_0^2 N^3 D^5}{Q^{0.56}}\right)^{0.45} \tag{25}$$

with a standard deviation of 3 %, where c is a constant and P_0 is the power consumption without gas induction. Equation (25) takes the form suggested by Michel and Miller [61] for aerated and agitated conventional gas–liquid contactors. Kasundra et al. correlated power consumption with the speed difference with N_C:

$$\frac{P}{P_0} = 1 - c(N - N_C)^b \tag{26}$$

where b and c are empirical constants, which were estimated for single gas-inducing impeller and that coupled with another impeller for gas–liquid mixture circulation [23]. Poncin et al. [41] correlated gassed power number as a function of $(Fr - Fr_C)$ for a hollow disk impeller:

$$N_P = 0.19 + \frac{0.92}{1 + 0.57(Fr - Fr_C)^2} \tag{27}$$

where Fr_C is the Froude number at critical impeller speed. For the impeller of his turbo aerator, Zundelevich [32] derived the following equation:

$$\frac{Qh}{P} = \frac{1}{N_P \rho g Fr}\left(\frac{ND^3}{Q_L}\frac{1}{EuFr} + \sqrt{\frac{1}{EuFr}}\right) \tag{28}$$

where Eu is the gas Euler number $[gh/(Q/D^2)^2]$ and Q_L is the liquid flow rate. Saravanan et al. [38] investigated the power consumption in gas-inducing reactors equipped with single and multiple impellers with reactor diameters of 0.57, 1.0, and 1.5 m. The correlation is as follows:

$$\frac{P - 2\pi\tau_{rg}N}{\rho W N^3 R^4} = C_{DO} - C_{DY}\left(1 - \frac{1}{\phi F}\right) \tag{29}$$

where W is the blade width, R the radius of impeller, τ_{rg} the constant torque acting on the eye of impeller, ϕ the vortex constant, and C_{DO} and C_{DY} are constants. These parameters were estimated for single- and dual-impeller systems with satisfactory results. Although various empirical correlations of power requirement have been obtained, the validity of the equations should be checked before application.

4.4 Gas–Liquid Mass Transfer Coefficient

The resistance to mass transfer across the gas–liquid interface is mainly caused by the liquid film. The correlation of lumped volumetric mass transfer coefficient, $k_L a$, where k_L is the liquid side mass transfer coefficient and a is the specific gas-liquid interfacial area, with operating variables and liquid properties, has been extensively studied and reviewed for conventional aerated and agitated gas–liquid contactors [2, 62]. The correlations are empirical, and typically, they take the form [63, 64]:

$$k_L a = c\left(\frac{P}{V}\right)^{\alpha} v_S^{\beta} \tag{30}$$

or [65, 66]

$$k_L a = cN^{\alpha} v_S^{\beta} \tag{31}$$

where c, α, and β are constants; V is liquid volume; P is power consumption; N is rotation speed; and v_S is superficial gas velocity. The liquid properties also exert influences on $k_L a$. For conventional gas liquid contactors, Robinson and Wilke [67] obtained the correlation

$$k_L a = c\left(\frac{P}{V}\right)^{\alpha 1} v_S^{\alpha 2} \rho^{\alpha 3} D_L^{\alpha 4} \sigma^{\alpha 5} \mu^{\alpha 6} \tag{32}$$

where $\alpha 1$ through $\alpha 6$ are constants, D_L is the gas diffusivity in liquid, σ is the surface tension, and μ is the liquid viscosity. Dimensional analysis gives:

$$\frac{k_L a D^2}{D_L} = c\left(\frac{\rho N D^2}{\mu}\right)^{\alpha 1}\left(\frac{DN^2}{g}\right)^{\alpha 2}\left(\frac{\mu}{\rho D_L}\right)^{\alpha 3}\left(\frac{\mu v_S}{\sigma}\right)^{\alpha 4}\left(\frac{ND}{v_S}\right)^{\alpha 5} \tag{33}$$

or

$$Sh = c \cdot Re^{\alpha 1} Fr^{\alpha 2} Sc^{\alpha 3} \left(\frac{\mu v_S}{\sigma}\right)^{\alpha 4} Ae^{\alpha 5} \tag{34}$$

where *Sh*, *Re*, *Fr*, *Sc*, and *Ae* are Sherwood number, Reynolds number, Schmidt number, and aeration number, respectively. Yagi and Yoshida [68] used this type of correlation to estimate the $k_L a$ for Newtonian and non-Newtonian fluids.

For gas self-inducing reactors, correlations of $k_L a$ take the similar forms for conventional agitated and aerated reactors (e.g., Sawant et al. [31]). As mentioned above, a prominent difference between gas-inducing reactor and conventional reactor is the gas flow rate, which is an independent operating variable for the latter, while is dependent on the impeller speed for the former. At the critical speed, no gas flow occurs. Therefore, some correlations include variables at the critical speed. For example, Kasundra et al. [23] correlated $k_L a$ with the difference between power consumption under the operating condition and at the critical impeller speed (Table 1). Recently, Yu and Tan [69] proposed to add a term of impeller submergence in Eqs. (30) and (31), because the impeller submergence exerted significant influence on the gas-induction rate and $k_L a$ (Table 1). These equations could correlate the experimental data of Forrester et al. (concave six-blade hollow impeller) [17], Poncin et al. (two-disk hollow impeller) [41], and Kasundra et al. (a hollow gas-inducing impeller with a bottom impeller) [23]. These equations also well correlated their data obtained from a laboratory scale square reactor with a four-blade turbine inside a draft tube. Many studies have been carried out to investigate the characteristics of mass transfer in gas-inducing reactors, and the published correlations of $k_L a$ are summarized in Table 1.

The liquid properties also exert influences on mass transfer in gas self-inducing reactors. Therefore, many researchers have correlated $k_L a$ with operating variables and liquid properties. For instance, Heim et al. [14], Hichri et al. [70], and Patil et al. [71] obtained correlations in the form of non-dimensional numbers (Table 1).

The bubble diameter determines the specific area a at the same gas holdup. Girgin et al. [72] investigated the bubble size in a laboratory Denver flotation cell with a volume of 5.5 L by using a McGill bubble size analyzer. They observed that the bubble diameter increased with the increase in agitation speed when no or little frother (5 ppm) was added to the liquid phase. This finding is surprising because in conventional gas-liquid reactors, the bubble diameter usually decreases with increasing agitation speed. Since the gas-induction rate increases with the impeller speed, this phenomenon implies that the effect of increased gas-induction rate is more profound than that of increased shear caused by agitation. The change of bubble diameter was not appreciable when 20 or 50 ppm frother was supplied due to reduced bubble coalescence. The gas-induction rate decreased with increased frother concentration at high rotation speed, because reduced bubble diameter means less buoyancy, higher gas holdup, lower fluid density, and power consumption. Gas holdup ε influences the specific interfacial gas–liquid area a and

Table 1 Correlation of $k_L a$ in gas self-inducing reactors

Reactor	Geometry	Gas/liquid system	Correlation	Reference
Hollow shaft with hollow pipe or flattened cylinder	$D = 0.41$, 0.57 and 1.0 m	Air/alkaline solution of sodium dithionate	$k_L a = 6.8 \times 10^{-3} \cdot (P/V)^{0.55} v_S^{0.5}$ for $v_S < 0.005$ m/s $k_L a = 3.26 \times 10^{-3} \cdot (P/V)^{0.55} v_S^{0.25}$ for $v_S > 0.005$ m/s $a = 112 \cdot (P/V)^{0.4} v_S^{0.5}$ for $v_S < 0.005$ m/s $a = 36.7 \cdot (P/V)^{0.4} v_S^{0.25}$ for $v_S > 0.005$ m/s $\varepsilon = 1.12 \times 10^{-3} \cdot (P/V)^{0.5}$	[12]
Denver-type flotation cell	100 × 100, 140 × 140, 172 × 172, 380 × 380 mm	Air/sodium dithionite CO_2/ $Na_2CO_3 + NaHCO_3$	$k_L a = 0.0195 \cdot (P/V)^{0.5}$ $a = 75 \cdot (P/V)^{0.5}$ $\varepsilon = 0.0325 \cdot (P/V)^{0.5}$	[31]
Hollow shaft with hollow tube and 3-edge impellers	$V = 6.3 \sim 780$ L	O_2/aqueous hydrazine solution	$k_L a = 1.1 \times 10^{-4} \cdot (P/V)^{0.8} (\mu/\rho g^2)^{-1/3}$	[73]
Hollow shaft and hollow disk turbine	$T = 100$ mm	H_2/tetralin, hydrogenated coal-derived liquid	$k_L a = 3.43 \times 10^{-4} \cdot (P/V)^{0.8} (H/T)^{-1.9}$	[74]
Hollow shaft	Liquid volume 3.954 L	H_2, N_2, CH_4/ n-hexane	$Sh = 6.67 \times 10^{26} \cdot Eu^{-1.0} We^{4.2} Sc^{0.1} Re^{-2.6}$	[75]
Hollow shaft	Liquid volume 3.954 L	N_2, CH_4/H_2O, $n\text{-}C_6H_{14}$	$Sh = 2.39 \times 10^{-28} \cdot Re^{4.86} Sc^{2.55} We^{0.34} Eu^{0.35}$	[75]
Hollow shaft	$D = 0.29$ m, volume 150 L	H_2/2-propanol, *o*-cresol	$Sh = 0.123 \cdot Re^{0.44} Sc^{0.5} We^{1.57} (V_G/V_L)^{1.1}$	[70]
Hollow shaft	$D = 0.29$ m, volume 150 L	H_2/*o*-cresol	$Sh = 5 \times 10^{-03} \cdot Re^{0.717} Sc^{0.5} We^{1.17}$	[76]

(continued)

Table 1 (continued)

Reactor	Geometry	Gas/liquid system	Correlation	Reference
Hollow shaft and hollow disk turbine	$H/D = 1$ $H/D = 1.4$	H_2/H_2O, ethanol, adiponitrile	$Sh = 3 \times 10^{-4} \cdot Re^{1.45} Sc^{0.5} We^{0.5}$ $Sh = 1.5 \times 10^{-4} \cdot Re^{1.45} Sc^{0.5} We^{0.5}$	[77]
Hollow shaft		N_2, H_2, CH_4, CO, CO_2/methanol, ethanol	$Sh = 4.88 \times 10^{6} \cdot Re^{-3.81} Sc^{0.23} We^{4.48} Eu^{0.09}$	[78]
Hollow shaft 4-pipe	$D = 300$ mm, $H = 255 \sim 455$ mm	Air/H_2O, fermentation broth with mycelia	$Sh = 9.5 \times 10^{-5} \cdot \left[1 - \exp\left(-19.94 \cdot Re^{-0.216} Fr^{1.336}\right)\right]$ $\varepsilon = 28.96 \cdot Fr^{0.968} Ae^{0.644} Re^{-0.354} (D/h)^{0.176}$	[14]
Hollow shaft 6-pipe	$D = 300$ mm, $H = 255 \sim 455$ mm	Air/H_2O, fermentation broth with mycelia	$Sh = 1.06 \times 10^{-4} \cdot [1 - \exp(-21.63 \cdot Re^{-0.234} Fr^{1.207})]$ $\varepsilon = 199.7 \cdot Fr^{1.063} Ae^{0.789} Re^{-0.458} (D/h)^{0.316}$	[14]
Hollow shaft ratchet	$D = 300$ mm, $H = 255 \sim 455$ mm	Air/H_2O, fermentation broth with mycelia	$Sh = 1.04 \times 10^{-4} \cdot \left[1 - \exp\left(-1331.2 \cdot Re^{-0.557} Fr^{2.498}\right)\right]$ $\varepsilon = 25.85 \cdot Fr^{0.947} Ae^{0.634} Re^{-0.336} (D/h)^{0.255}$	[14]
Standpipe with two turbines of different types	$D = 0.19 \sim 0.5$ m, $T = 0.57 \sim 1.5$ m	Air/H_2O	$\varepsilon = A \cdot (P/V)^{b} v_{S}^{c}$ $A = 0.00246 \sim 0.00283$, $b = 1.32 \sim 2.01$, $c = 0.37 \sim 0.52$ for different turbine combinations	[37]
Hollow shaft, hollow 6-blade impeller	$D = 127$ mm	O_2, N_2/cyclohexane	$Sh = 4.51 \times 10^{3} \cdot We^{-0.21} Fr^{0.92} (1 + 1.867 \times 10^{3} \varepsilon)$ $\varepsilon = 0.0385 \cdot \left(\frac{N-11.6}{11.6}\right)^{1.16} (\mu_G/\mu_L)^{-0.74} (\rho_G/\rho_L)^{-0.82} (\sigma/0.025)^{1.97}$	[79]
Hollow shaft, hollow 6-blade impeller	$D = 114$ mm	O_2, N_2/cyclohexane	$\ln(k_L a) = \beta_0 + \sum_{i=1}^{4} \beta_i x_i + \sum_{i=1}^{4} \beta_{ii} x_i^2 + \alpha_1 \exp\left(\gamma_1 (x_1 + \gamma_2)^2\right) + \alpha_2 \exp(\gamma_3 (x_1 + 3)(4 - x_4))$ x_1, coded N; x_2, coded pressure; x_3, coded temperature; x_4, coded H	[81]

(continued)

Table 1 (continued)

Reactor	Geometry	Gas/liquid system	Correlation	Reference
Standpipe with two turbines and a sparger	$D = 0.5$ m, $T = 1.5$ m	Air/H_2O	$\varepsilon = A \cdot (\mathrm{Re}FrFl)^b (C_3D)^c$ $A = 0.109 \sim 0.221$, $b = 0.375 \sim 0.433$, $c = -0.132 \sim 0.334$ for different spargers, Fl: flow number	[39]
Hollow shaft and six concave blades with multiple holes	$T = H = 0.45$ m	Air/water	$k_La = 76 \times 10^6 \cdot (P/V)^{0.8}$	[57]
Zipperclave with a hollow shaft and a hollow pipe impeller	4 L, $T = 114$ mm	H_2, N_2/soybean oil	For N_2 $\ln(k_La) = -4.86 - 0.179x_1 + 0.708x_2 - 0.596x_3 + 0.0759x_1^2 + 0.116x_2^2 - 0.228x_1x_2 - 0.0763x_2x_3 - 0.0754x_1^3 + 0.00269(x_2 + 2.5)\exp(2x_3) + 1.28\tanh(0.3x_2(5.5 - x_3^2) + 0.1(2 - 4x_3)) - 0.339x_1x_2x_3$ For H_2 $\ln(k_La) = -3.868 + 0.516x_2 - 0.790x_3 + 0.223x_1^2 - 0.352\exp(x_1) + 0.326\exp(x_3) - 0.00378(x_2 + 3)\exp(2.5x_3) + 2.099\tanh(0.3x_2(8 - x_3^2) + 0.1(2 - 6x_3)) - 0.927x_1\exp(-\lvert x_2\rvert)$ x_1, x_2, and x_3 are coded variables for temperature, impeller speed, and liquid height in reactor	[82]
Hollow shaft with hollow vaned disk	$T = 0.6$ m	Air/potassium sulfate	$k_La = \frac{0.065(Fr - Fr_C)^{1.1}}{1 + 0.132(Fr - Fr_C)^{1.1}}$ for coalescing system $k_La = \frac{0.35(Fr - Fr_C)^{1.6}}{1 + 0.2(Fr - Fr_C)^{1.6}}$ for non-coalescing system $\varepsilon = \frac{(Fr - Fr_C)^{1.1}}{0.043(Fr - Fr_C)^{1.1} + 0.135}$	[41]

(continued)

Table 1 (continued)

Reactor	Geometry	Gas/liquid system	Correlation	Reference
Hollow shaft and impellers	$T = 0.5, 1.0$ m	Air/hydrazine	$k_L a = c\left[(P/V) - (P/V)_C\right]^{0.8}$ $c = 0.0243 \sim 0.0772$ for different impeller combinations	[23]
Stator–rotor with a standpipe	$T = 1.0, 1.5$ m, 8.2×8.2 m	Air/sodium sulfite	$k_L a = 4.97 \times 10^{-3} \cdot (D^3/V)(\mu/\rho g^2)^{-1/3} Fr^{0.56} (V_A/V)^{0.32}$ V_A, active volume	[71]
Hollow shaft with a hollow disk turbine	$T = 64$ mm, $D = 32$ mm	N_2, H_2/water, n-hexadecane, toluene, acetone, sunflower oil	$Sh = 10^{-4.41} Re^{1.66} Sc^{1.08}$	[80]
Four-blade turbine with a draft tube inside a square reactor	$T = 0.406$ m	N_2/water	$k_L a = 1.212 \cdot (P/V)^{0.0816} v_S^{0.692} (S/T)^{-0.390}$ $k_L a = 2.504 \cdot (N/N_{cd})^{-2.20} v_S^{0.691} (S/T)^{-0.630}$ N_{cd}, the minimum impeller speed for complete dispersion of gas phase	[69]

thus affects k_La. Some researchers correlated the gas holdup in hollow pipe and Denver-type gas-inducing reactors, and the correlations are also shown in Table 1.

4.5 Performance Prediction and Simulation

In addition to the above-mentioned mechanistic and empirical models, other approaches have also been used to understand the performance of gas-inducing reactors. The statistical design and analysis method is an effective way to understand the influences of multiple variables on the process performances through designed, less number of experiments and has been widely applied for process optimization. Tekie et al. [81] used a central composite design and analysis approach to study the mass transfer of oxygen and nitrogen in cyclohexane and correlated k_La with agitation speed, pressure, temperature, and submergence of a gas self-inducing reactor (Table 1). They concluded that their statistical model was more accurate than the dimensionless correlations reported in the literature. Later, Fillion and Morsi [82] investigated the transfer of nitrogen and hydrogen in soybean oil using the similar method. They found that pressure did not affect k_La and obtained a correlation of k_La with temperature, rotation speed, and submergence (Table 1).

Artificial neural network has been used to correlate the mass transfer performance with the geometrical, operating, and physicochemical variables. For example, Reisener et al. [83] studied mass transfer in electrolytic airlift and bubble column reactors using artificial neural network. The trained network could model the mass transfer characteristics not only for the reactors they used, but also for the reactors in other studies. Yang et al. [84] used neural network to model k_La using data from different publications. Their neural network was consisted of an input layer for six variables (power consumption per unit volume, superficial gas velocity, liquid height/reactor diameter ratio, impeller diameter/reactor diameter ratio, aeration rate, and liquid viscosity), a hidden layer with 11 nodes, and an output node k_La. The trained network could predict the mass transfer coefficients of different authors with standard error of ±25 %. Lemoine et al. [85] established a database with 4435 data points from the literature, and the neural network that contained an input layer, two hidden layers, and an output layer could predict k_La in both surface aeration and gas-inducing reactors. They further constructed a database including 7374 data points, and the artificial neural network could predict gas holdup, critical rotation speed, Sauter mean bubble diameter, specific area, and mass transfer coefficient of gas-inducing reactors, surface aeration reactors, and gas sparging reactors with improved precision [86].

With the progresses of fluid mechanics and computer technology, computational fluid dynamics (CFD) is widely applied for understanding the behavior and performance of liquid flow in reactors. CFD is a powerful tool for analysis of the distribution of liquid and gas flow, and gas holdup. Nowadays, CFD has been used to predict critical rotation speed, gas-induction rate, mass transfer, dynamic behavior of cell growth and metabolite production and is useful to understand the characteristics and weaknesses of a self-inducing reactor to further improve its performances. In 1998, Rigby and Evans studied the pressure coefficient C_P of a self-inducing cylindrical impeller using CFD [87]. Murthy et al. [88] simulated a self-inducing reactor equipped with a hollow shaft and a hollow pitched-blade impeller with multiple holes on each blade, and the simulation was based on conservation equations for mass and momentum for each phase. Simulation indicated that for single-phase system (no gas induction occur) and two-phase system, all the low-pressure zones appeared on the top edge of blades. Experiments showed that the gas-induction rate produced by the impeller with gas egress holes on the upper side of blades was 100 $\sim$ 200 % higher than that with holes on the lower side of blades. The calculated gas-induction rate was accurate at low impeller speed, but was slightly underestimated at high impeller speed. This study provides a good example of improvement of the gas-induction performance of gas-inducing impeller by using CFD. Achouri et al. [89] used CFD to calculate the gas-induction rate, power consumption, and power number of a reactor with a downflow pitched-blade impeller with blade angle of 45° or 60°. Their calculations indicated that the PBTD 60° impeller was more efficient than the PBTD 45° impeller under the same operating condition, because the gas-induction rate of the former was higher, although the latter showed less power consumption. Recently, Fonte et al. [90] investigated a gas-inducing reactor with a six-blade turbine on a hollow shaft using CFD. The ratio of blade width to turbine diameter was as high as 1.0. Single liquid phase simulation was carried out, and the pressure difference between the gas inlets and outlets on the hollow shaft was correlated with the impeller speed. The gas-induction rate could correlate to the pressure loss inside the hollow shaft as well as the impeller speed at low gas holdup.

For gas–liquid–solid dispersion in self-inducing reactors, mixing should be extensive enough to suspend the solid particles. Murthy et al. [91] investigated gas–liquid–solid dispersion using CFD in reactors equipped with different gas-inducing turbines. The predicted distribution of gas holdup well agreed with the results of cone beam X-ray computed tomography (CBCT). CFD has been used to study on the collision between gas bubbles and solid particles in flotation cells by Koh et al. The investigated flotation cells included Outokumpu cell [92], Denver cell [93, 94], and flotation cells developed at CSIRO [95, 96] and at Mitso Minerals [92]. Simulation indicated that in the CSIRO flotation cell, the distribution of turbulent dissipation rate and bubble–particle collision rate was much better than conventional reactor equipped with a Rushton turbine [95]. In a Denver flotation cell, the predicted gas-induction rates agreed well with the measured rates at frother concentrations of 20 and 50 ppm, but underestimated at zero or low frother concentrations [94], and the behavior was consistent with that reported by Girgin et al. [72]. The results of attachment and net detachment of coarse and fine particles confirmed the reasonability of the practice to treat coarse and fine particles streams separately in many flotation plants.

Recently, Hong et al. [97] simulated the liquid flow in a reactor equipped with a six-blade ratchet, self-inducing turbine coupled with a 12-channel stator. The CFD model was coupled with a kinetic model of yeast growth limited by the sugar concentration. The calculated profiles of cell density, sugar concentration, and DO were consistent with the measured results.

Industrial tomography is a powerful tool to investigate the behavior of gas–liquid and gas–liquid–solid dispersion in reactors and has been used for the study on gas self-inducing reactors. Vesselinov et al. [98] carried out CFD simulation of a self-inducing reactor and compared with the results obtained from cone beam computed tomography (CBCT). Boden et al. [99]. and Hampel et al. [100]. studied the gas holdup distribution in a gas-inducing reactor by using X-ray and ^{137}Cs γ-ray CBCT.

5 Applications in Biotechnology

The self-inducing reactors are widely applied in biological treatment of wastewater and fermentation, typically in vinegar production [101] using Frings Acetator. In China, the self-inducing-type bioreactors were used in industrial fermentation processes in the later 1970s. The fermentor capacity was 13 m^3 for vinegar, 20 m^3 for rifamycin SV, 50 m^3 for oxytetracycline, and 50 m^3 for yeast cell production. Pilot gas self-inducing fermenters were also used for yeast production from wastewater of glutamate production industry and glucoamylase production. These self-inducing fermenters were operated in the mode of sucking air from outside of fermentor without using an air compressor.

High cell-density culture is widely carried out for recombinant protein production to enhance the volumetric productivity and to reduce the investment cost. With the increase in cell density, the oxygen demand increases as well, and pure oxygen is used to improve oxygen supply. However, the oxygen utilization efficiency is poor due to its low solubility. Matsumura et al. proposed to enhance the oxygen utilization by gas entrainment from the headspace of the fermentor [47]. The dead-end operating mode may be not suitable because accumulated CO_2 can bring about negative effects, and oxygen recycle with a suitable ventilation rate can result in better efficiency of oxygen utilization and avoiding accumulation of waste gaseous products.

Facing the situation of global consumption of tremendous amounts of fossil energy sources together with emission of huge amount of CO_2, a greenhouse gas causing the global temperature rise, investigations and practices on production of fuels and chemicals from renewable resources have been carried out worldwide. Some bacteria can use CO and CO_2 as the carbon source and H_2 as the electron donor for production of biofuels and chemicals [102, 103]. Production of fuels and chemicals from flue gas or synthesis gas (syngas) derived from gasification of lignocellulosic biomass is promising because such fermentation does not impact food supply. Since the solubility of CO and H_2 is very low, the fermentation process is limited by the gas–liquid mass transfer. Conventional agitated bioreactors [104] and hollow fiber membrane bioreactors [105] have been used for the fermentation, but the k_La of the latter was shown to be the lowest among the reactors examined [106]. CO and H_2 are flammable, the former is toxic, and their solubilities are very low. Therefore, use of gas-inducing reactor operated in the dead-end mode is expected to be favorable to efficient gas substrate utilization.

In some kind of metabolite production, CO_2 can be a substrate. For example, anaerobic production of succinic acid from glucose theoretically requires 0.58 mol CO_2 per mol succinate produced [107]. Use of traditional gassed and agitated reactors results in emission of CO_2 and low CO_2 utilization efficiency. A self-inducing reactor operated in the dead-end mode was used in succinic acid fermentation to fully use the supplied CO_2 gas, and no CO_2 was released [27]. In the process of ethanol fermentation with glucose as the substrate, 1 mol CO_2 is formed when 1 mol ethanol is produced and CO_2 is a main byproduct. To reduce CO_2 emission in ethanol fermentation, succinic acid fermentation (*Escherichia coli* NZN 111) in a self-inducing fermentor was further coupled with ethanol fermentation to completely use the CO_2 formed in ethanol production (Fig. 6).

Microalgae grow through photosynthesis with CO_2 as the carbon source, and some strains can accumulate a large quantity of oil which can be converted to biofuel. Compared to the process of succinic acid fermentation, cultivation of algae consumes much more CO_2 based on formed target metabolite, and self-inducing bioreactors are expected to be useful in the culture process with CO_2 or flue gas as the carbon source. Kshirsagar et al. [108] carried out batch culture of *Spirulina platensis* in a laboratory bioreactor equipped with a self-inducing impeller. The CO_2 contained in the atmosphere was used as the carbon source. Since the CO_2 content in air is very low, the bioreactor was not operated in the dead-end mode. The cell density reached 4.44 g/L in 15 days, while in shake flask culture, it took 25 days to reach a cell density of 3.56 g/L, indicating improved CO_2 supply in the self-inducing bioreactor. However, cultivation of algae in gas-inducing reactors operated in the dead-end mode has not been reported.

For high-throughput screening, miniaturized multiple bioreactor systems have been developed. Weuster-Botz's team has established a system comprising 48 bioreactors with volume of 10 ml each. The reactors were capable of at-line pH and

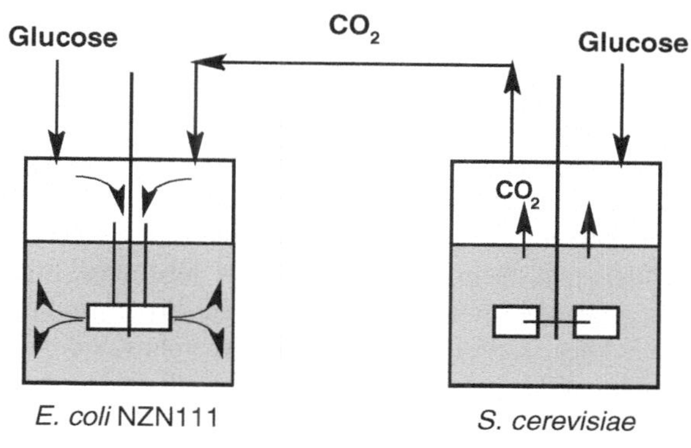

Fig. 6 Coupling a succinate fermentation process with ethanol fermentation to utilize the CO_2 produced by *Saccharomyces cerevisiae*

dissolved oxygen (DO) measurement and were agitated with self-inducing impellers. The cell density of *E. coli* achieved 16.5 [109], 20.5 [110], and 36.9 g/L [111], much higher than that obtained in microtiter plates. The $k_L a$ of the miniature bioreactors could reach a level as high as 0.4 s^{-1} [110]. The miniature bioreactor system was also used for scale-down of riboflavin production by a recombinant *Bacillus subtilis* strain, and the riboflavin titer was similar to that obtained in a 7.5-L lab bioreactor [112]. Furthermore, the reproducibility in the miniature bioreactors for cultivation of *E. coli* [113] or *B. subtilis* [114] was satisfactory. This miniature bioreactor system has also been used to compare the production of 2-hydroxyisobutyric acid (2-HIBA) by recombinant strains of *Cupriavidus necator*, and the effects of pH and fructose feeding on 2-HIBA production in the miniaturized reactors were studied [115]. Fed-batch and two-stage cultures were performed in a 2-L agitated bioreactor, and the results showed that the concept derived from the milliliter reactors was feasible on a liter-scale reactor, indicating the miniature bioreactor system is a powerful tool for strain screening and culture process improvement.

Animal cells are shear sensitive and can be damaged by vigorous agitation. The bioreactor equipped with a helical ribbon impeller and vertical surface baffles developed by Kamen et al. [54] is suitable to cultivation of shear-sensitive cells due to low shear. Reactors of this type were successfully applied to culture of shear-sensitive animal cell (insect cell sf-9) [116] and plant cell (*Catharanthus roseus*) [117].

Although the gas self-inducing reactor has been applied in biotechnology, the reported cases of application to modern biotechnology are limited as compared to the use of conventional gas–liquid reactors. As mentioned above, its prominent advantage is high efficiency of gas utilization. With the development of white biotechnology, gas substrates such as CO and CO_2 have the potential for production of biofuels and biochemicals by microorganisms, and the gas self-inducing bioreactors are expected to play an important role when the processes are commercially scaled up.

6 Conclusion

A gas self-inducing impeller can recycle the gas phase from the headspace of a gas–liquid contactor without additional equipment, such as a gas compressor, and thus effectively enhances the gas utilization efficiency and reduces investment cost. This feature is favorable not only to chemical reactions, but also to biochemical systems. Self-inducing bioreactors are expected to be applied to bioprocesses for production of fuels and chemicals from gaseous substrates, high-throughput screening microbial strains, fermentation process scale–down, and so forth. Although many studies have been carried out, further researches are still required for better design, understanding, scale-up and commercialization of gas self-inducing reactors.

References

1. Garcia-Ochoa F, Gomez E (2009) Bioreactor scale-up and oxygen transfer rate in microbial processes: An overview. Biotechnol Adv 27:153–176
2. Joshi JB, Pandit AB, Sharma MM (1982) Mechanically agitated gas-liquid reactors. Chem Eng Sci 37:813–844
3. Hromatka O, Ebner H (1959) Vinegar by submerged oxidative fermentation. Ind Eng Chem 51:1279–1280
4. Matsumura M, Sakuma H, Yamagata T, Kobayashi J (1982) Gas entrainment in a new gas entraining fermentor. J Fermen Technol 60:457–467
5. Patwardhan AW, Joshi JB (1999) Design of gas-inducing reactors. Ind Eng Chem Res 38:49–80
6. Long WH, Castle AW (1877) Improvement in mercury-condensers. US Patent 191,805
7. Ruth Jr JP (1918) Aeration device. US Patent 1,268,630
8. Fulweiler WH, Jordan CW (1926) Process of purifying gas. US Patent 1,632,758
9. Enenkel A, Maurer R (1955) Apparatus for the aeration of liquids. British Patent 724,791
10. Ebner H, Enenkel A (1972) Device for aerating liquid. US Patent 3,813,086
11. Evans GM, Rielly CD, Davidson JF, Carpenter KJ (1992) Hydrodynamic characteristics of a gas-inducing impeller. Fluid Mechan Mix 10:153–161
12. Joshi JB, Sharma MM (1977) Mass transfer and hydrodynamic characteristics of gas inducing type of agitated contactors. Can J Chem Eng 55:683–695
13. Zlokarnik M, Judat H (1967) Tube and disk stirrers—two efficient stirrers for the gassing of liquids. Chem Ing Tech 39:1163–1168
14. Heim A, Kraslawski A, Rzyski E, Stelmach J (1995) Aeration of bioreactors by self-aspirating impellers. Chem Eng J 58:59–63
15. Forrester SE, Rielly CD (1994) Modelling the increased gas capacity of self-inducing impellers. Chem Eng Sci 49:5709–5718
16. Rielly CD, Evans GM, Davidson JF, Carpenter KJ (1992) Effect of vessel scaleup on the hydrodynamics of a self-aerating concave blade impeller. Chem Eng Sci 47:3395–3402
17. Forrester SE, Rielly CD, Carpenter KJ (1998) Gas-inducing impeller design and performance characteristics. Chem Eng Sci 53:603–615
18. Martin GQ (1972) Gas-inducing agitator. Ind Eng Chem Proc Des Dev 11:397–404
19. Deshmukh NA, Path SS, Joshi JB (2006) Gas induction characteristics of hollow self-inducing impeller. Chem Eng Res Des 84(A2):124–132
20. Zlokarnik M (1966) Gasdurchsatzes zur Flüssigkeitsbegasung Bestimmung des Gasdurchsatzes und der Wellenleistung. Chem Ing Tech 38:357–366
21. Achouri R, Hamza SB, Dhaouadi H, Mhiri H, Bournot P (2013) Volumetric mass transfer coefficient and hydrodynamic study of a new self-inducing turbine. Energy Conv Manag 71:69–75
22. Mohammed AK, Hussen HA, Al-Rassul SA (2008) Performance of gas induction in a dual-impeller agitated bioreactor. Al-Khwarizmi Eng J 4(4):1–8
23. Kasundra RB, Kulkarni AV, Joshi JB (2008) Hydrodynamic and mass transfer characteristics of single and multiple impeller hollow self-inducing reactors. Ind Eng Chem Res 47:2829–2841
24. Ju F, Cheng ZM, Chen JH, Chu XH, Zhou ZM, Yuan PQ (2009) A novel design for a gas-inducing impeller at lowest critical speed. Chem Eng Res Des 87:1069–1074
25. Gomadurai C, Saravanan K, Abraham E, Deepa N (2014) Hydrodynamic studies on air-inducing impeller system. Int J ChemTech Res 6:4471–4474
26. Mundale VD, Joshi JB (1995) Optimization of impeller design for gas inducing type mechanically agitated contactors. Can J Chem Eng 73:161–172
27. Wu H, Li Q, Li Z, Ye Q (2012) Succinic acid production and CO_2 fixation using a metabolically engineered *Escherichia coli* in a bioreactor equipped with a self-inducing agitator. Biores Technol 107:376–384

28. Joshi JB (1980) Modifications in the design of gas inducing impeller. Chem Eng Comm 5:109–114
29. Wang Z, Xu P, Li X, Wang S, Cheng Z, Ju F (2013) Impact of liquid driving flow on the performance of a gas-inducing impeller. Chem Eng Process 69:63–69
30. Sawant SB, Joshi JB (1979) Critical impeller speed for the onset of gas-inducing types of agitated contactors. Chem Eng J 18:87–91
31. Sawant SB, Joshi JB, Pangarkar VG, Mhaskar RD (1981) Mass transfer and hydrodynamic characteristics of the Denver type of flotation cells. Chem Eng J 21:11–19
32. Zundelevich Y (1979) Power consumption and gas capacity of self-inducing turbo aerators. AIChE J 25:763–773
33. Raidoo AD, Rao KSMSR, Sawant SB, Joshi JB (1987) Improvements in gas inducing impeller desigh. Chem Eng Commun 54:241–264
34. Mundale VD, Joshi JB (1995) Optimization of impeller design for gas inducing type of agitated contactor. Can J Chem Eng 73:6–17
35. Saravanan K, Mundale VD, Joshi JB (1994) Gas inducing type mechanically agitated contactors. Ind Eng Chem Res 33:2226–2241
36. Saravanan K, Joshi JB (1995) Gas-inducing-type mechanically agitated contactors: hydrodynamic characteristics of multiple impellers. Ind Eng Chem Res 34:2499–2514
37. Saravanan K, Joshi JB (1996) Fractional gas hold-up in gas inducing type of mechanically agitated contactors. Can J Chem Eng 74:16–30
38. Saravanan K, Mundale VD, Patwardhan AW, Joshi JB (1996) Power consumption in gas-inducing-type mechanically agitated contactors. Ind Eng Chem Res 35:1583–1602
39. Patwardhan AW, Joshi JB (1997) Hydrodynamics of a stirred vessel equipped with a gas-inducing impeller. Ind Eng Chem Res 36:3904–3914
40. Patil SS, Joshi JB (1999) Stability of gas-inducing type impellers. Can J Chem Eng 77:793–803
41. Poncin S, Nguyen C, Midoux N, Breysse J (2002) Hydrodynamics and volumetric gas-liquid mass transfer coefficient of a stirred vessel equipped with a gas-inducing impeller. Chem Eng Sci 57:3299–3306
42. Sardeing R, Poux M, Melen S, Avrillier P, Xuereb C (2005) Aeration of large size tanks by a surface agitator. Chem Eng Technol 28:587–595
43. Sardeing R, Ferrand F, Poux M, Avrillier P, Xuereb C (2003) Hydrodynamics and gas dispersion characterization in a system equipped with a new gas-inducing impeller. Eng Life Sci 3:31–37
44. Sardeing R, Xuereb C, Poux M (2006) Improvement of the performances of a gas-inducing system for application in wastewater treatment. Int J Chem React Eng 4:A30
45. Sardeing RF, Poux M, Xuereb C (2006) Development of a new gas-inducing turbine family: the partially shrouded turbine. Ind Eng Chem Res 45:4791–4804
46. Patwardhan AW, Joshi JB (1998) Design of stirred vessels with gas entrained from free liquid surface. Can J Chem Eng 76:339–364
47. Matsumura M, Sakuma H, Yamagata T, Kobayashi J (1980) Gas absorption characteristics of the tank-type fermentor with a mechanical gas entrainer. J Ferment Technol 58:69–77
48. Hsu YC, Chang H-C (1995) Onset of gas self-induction and power consumption after gas induction in an agitated tank. J Chem Tech Biotechnol 64:137–148
49. Hsu Y-C, Huang C-J (1996) Characteristics of a new gas-induced reactor. AIChE J 42:3146–3152
50. Hsu Y-C, Peng RY, Huang C-J (1997) Onset of gas induction, power consumption, gas hold up and mass transfer in a new gas-induced reactor. Chem Eng Sci 52:3883–3891
51. Chen J-H, Hsu Y-C, Chen Y-F, Lin C-C (2003) Application of gas-inducing reactor to obtain high oxygen dissolution in aeration process. Water Res 37:2919–2928
52. Scargiali F, Russo R, Grisafi F, Brucato A (2007) Mass transfer and hydrodynamic characteristics of a high aspect ratio self-ingesting reactor for gas-liquid operations. Chem Eng Sci 62:1376–1387

53. Scargiali F, Busciglio A, Grisafi F, Brucato A (2012) Gas-liquid-solid operation of a high aspect ratio self-ingesting reactor. Int J Chem React Eng 10:A27
54. Kamen AA, Chavarie C, André G, Archambault J (1992) Design parameters and performance of a surface baffled helical ribbon impeller bioreactor for the culture of shear sensitive cells. Chem Eng Sci 47:2375–2380
55. Conway K, Kyle A, Rielly CD (2002) Gas-liquid-solid operation of a vortex-ingesting stirred tank reactor. Chem Eng Res Des 80:839–845
56. Patil SS, Joshi JB (1999) Optimum design of stator-rotor assembly in gas inducing type mechanically agitated reactors. Chem Eng Commun 174:215–231
57. Evans GM, Rielly CD, Davidson JF, Carpenter KJ (1990) A fundamental study of gas-inducing impeller design. Fluid Mixing IV 121:137–152 (Institution of Chemical Engineers Symposium Series)
58. White DA, de Villiers JU (1977) Rates of induced aeration in agitated vessels. Chem Eng J 14:113–118
59. Saravanan K, Patwardhan AW, Joshi JB (1997) Critical impeller speed for solid suspension in gas inducing type mechanically agitated contactors. Can J Chem Eng 75:664–676
60. Patil SS, Mundale VD, Joshi JB (2005) Mechanism of gas induction in a self-inducing impeller. Ind Eng Chem Res 44:1322–1328
61. Michel BJ, Miller SA (1962) Power requirements of gas-liquid agitated systems. AIChE J 8:262–266
62. Sideman S, Hortacsu O, Fulton JW (1966) Mass transfer in gas-liquid contacting systems. Ind Eng Chem 58(7):32–47
63. Cooper CM, Fernstrom GA, Miller SA (1944) Gas-liquid contactors. Ind Eng Chem 36:504–509
64. Rushton JH, Gallagher JB, Oldshue JY (1956) Gas-liquid contacting with multiple mixing turbines. Chem Eng Prog 52:319–323
65. Yoshida F, Ikeda A, Imakawa S, Miura Y (1960) Oxygen absorption rates in stirred gas-liquid contactors. Ind Eng Chem 52:435–438
66. Hyman D, Van den Bogaerde JM (1960) Gas-liquid contacting in small bench-scale stirred reactors. Ind Eng Chem 52:751–753
67. Robinson CW, Wilke CR (1973) Oxygen absorption in stirred tanks: a correlation for ionic strength effects. Biotechnol Bioeng 15:755–782
68. Yagi H, Yoshida F (1975) Gas absorption by Newtonian and non-Newtonian fluids in sparged agitated vessels. Ind Eng Chem Proc Des Dev 14:488–493
69. Yu H, Tan Z (2012) New correlations of volumetric liquid-phase mass transfer coefficients in gas-inducing agitated tank reactors. Int J Chem Reactor Eng 10:A50
70. Hichri H, Accary A, Puaux JP, Andrieu J (1992) Gas-liquid mass transfer coefficients in a slurry batch reactor equipped with a self-gas-inducing agitator. Ind Eng Chem Res 31:1864–1867
71. Patil SS, Deshmukh NA, Joshi JB (2004) Mass transfer characteristics of surface aerators and gas-inducing impellers. Ind Eng Chem Res 43:2765–2774
72. Girgin EH, Do S, Gomez CO, Finch JA (2006) Bubble size as a function of impeller speed in a self-aerated laboratory flotation cell. Min Eng 19:201–203
73. Zlokarnik M (1978) Sorption characteristics for gas-liquid contacting in mixing vessels. Adv Biochem Eng 8:133–151
74. Kara M, Sung S, Klinzing GE, Chiang SH (1983) Hydrogen mass transfer in liquid hydrocarbons at elevated temperatures and pressures. Fuel 62:1492–1498
75. Chang M-Y, Morsi BI (1991) Mass transfer characteristics of gases in aqueous and organic liquids at elevated pressures and temperatures in agitated reactors. Chem Eng Sci 46:2639–2650
76. Hichri H, Accary A, Andrieu A (1991) Kinetics and slurry-type reactor modeling during catalytic hydrogenation of o-cresol on Ni/SiO_2. Chem Eng Proc 30:133–140
77. Dietrich E, Mathieu C, Delmas H, Jenck J (1992) Raney-nickel catalysed hydrogenations: gas–liquid mass transfer in gas-induced stirred slurry reactors. Chem Eng Sci 47:3597–3604

78. Koneripalli N, Tekie Z, Morsi BI, Chang M-Y (1994) Mass transfer characteristics of gases in methanol and ethanol under elevated pressure and temperature. Chem Eng J 54:63–77
79. Tekie Z, Li J, Morsi BI, Chang M-Y (1997) Gas–liquid mass transfer in cyclohexane oxidation process using gas-inducing and surface-aeration agitated reactors. Chem Eng Sci 52:1541–1551
80. Zieverink MMP, Kreutzer MT, Kapteijn F, Moulijn JA (2006) Gas-liquid mass transfer in bench scale stirred tanks—fluid properties and critical impeller speed for gas induction. Ind Eng Chem Res 45:4574–4581
81. Tekie Z, Li J, Morsi BI (1997) Mass transfer parameters of O_2 and N_2 in cyclohexane under elevated pressures and temperatures: a statistical approach. Ind Eng Chem Res 36:3879–3888
82. Fillion B, Morsi BI (2000) Gas-liquid mass-transfer and hydrodynamic parameters in a soybean oil hydrogenation process under industrial conditions. Ind Eng Chem Res 39:2157–2168
83. Reisener J, Reuter MA, Krűger J (1993) Modelling of the mass transfer in gas-sparged electrolysers with neural nets. Chem Eng Sci 48:1089–1101
84. Yang H, Fang BS, Reuss M (1999) $k_{L}a$ correlation established on the basis of a neural network model. Can J Chem Eng 77:838–843
85. Lemoine R, Fillion B, Behkish A, Smith AE, Morsi BI (2003) Prediction of the gas/liquid volumetric mass transfer coefficients in surface-aeration and gas-inducing reactors using neural networks. Chem Eng Process 42:621–643
86. Lemoine R, Morsi BI (2005) An algorithm for predicting the hydrodynamic and mass transfer parameters in agitated reactors. Chem Eng J 114:9–31
87. Rigby GD, Evans GM (1998) CFD simulation of gas dispersion dynamics in liquid cross-flows. Appl Math Model 22:799–810
88. Murthy BN, Deshmukh NA, Patwardhan AW, Joshi JB (2007) Hollow self-inducing impellers: flow visulization. Chem Eng Sci 62:3839–3848
89. Achouri R, Mokni I, Mhiri H, Bournot P (2012) A 3D CFD simulation of a self inducing pitched blade turbine dowflow. Energy Conv Manag 64:633–641
90. Fonte CP, Pinho BS, Santos-Moreau V, Lopes JCB (2014) Prediction of the induced gas flow rate from a self-inducing impeller with CFD. Chem Eng Technol 37:571–579
91. Murthy BN, Kasundra RB, Joshi JB (2008) Hollow self-inducing impellers for gas-liquid-solid dispersion: experimental and computational study. Chem Eng J 141:332–345
92. Koh PTL, Schwarz MP, Zhu Y, Bourke P, Peaker R, Franzidis JP (2003) Development of CFD models of mineral flotation cells. In: Proceeding of the third international conference on CFD in the minerals and process industries, Melbourne, Australia
93. Koh PTL, Schwarz MP (2003) CFD modelling of bubble–particle collision rates and efficiencies in a flotation cell. Min Eng 16:1055–1059
94. Koh PTL, Schwarz MP (2007) CFD model of a self-aerating flotation cell. Int J Min Proc 85:16–24
95. Koh PTL, Manickam M, Schwarz MP (2000) CFD simulation of bubble-particle collisions in mineral flotation cells. Min Eng 13:1455–1463
96. Koh PTL, Schwarz MP (2006) CFD modelling of bubble–particle attachments in flotation cells. Min Eng 19:619–626
97. Hong H-S, Cai Z-J, Li J-Q, Shi D-S, Wan W-Q, Li L (2014) Simulation of gas-inducing reactor couples gas-liquid mass transfer and biochemical reaction. Biochem Eng J 91:1–9
98. Vesselinov HH, Stephan B, Uwe H, Holger K, Günther H, Wilfried S (2008) A study on the two-phase flow in a stirred tank reactor agitated by a gas-inducing turbine. Chem Eng Res Des 86:75–81
99. Boden S, Bieberle M, Hampel U (2008) Quantitative measurement of gas hold-up distribution in a stirred chemical reactor using X-ray cone-beam computed tomography. Chem Eng J 139:351–362
100. Hampel U, Hristov HV, Bieberle A, Zippe C (2007) Application of high-resolution gamma ray tomography to the measurement of gas hold-up distributions in a stirred chemical reactor. Flow Meas Inst 18:184–190

101. García-García I, Santos-Dueñas IM, Jiménez-Ot C, Jiménez-Hornero JE, Bonilla-Venceslada JL (2009) Vinegar engineering. In: Solien L, Giudici P (eds) Vinegars of the world. Springer, Milan
102. Henstra AM, Sipma J, Rinzema A, Stams AJM (2007) Microbiology of synthesis gas fermentation for biofuel production. Curr Opin Biotechnol 18:200–206
103. Wilkins MR, Atiyeh HK (2011) Microbial production of ethanol from carbon monoxide. Curr Opin Biotechnol 22:326–330
104. Kundiyana DK, Huhnke RL, Wilkins MR (2010) Syngas fermentation in a 100-L pilot scale fermentor: design and process considerations. J Biosci Bioeng 109:492–498
105. Tsai S-P, Datta R, Basu R, Yoon S-H (2008) Syngas conversion system using asymmetric membrane and anaerobic microorganism. US Patent Application 12/036,007
106. Munasinghe PC, Khanal SK (2010) Syngas fermentation to biofuel: evaluation of carbon monoxide mass transfer coefficient ($k_L a$) in different reactor configurations. Biotechnol Prog 26:1616–1621
107. Vemuri GN, Eiteman MA, Altman E (2002) Effects of growth mode and pyruvate carboxylase on succinic acid production by metabolically engineered strains of *Escherichia coli*. Appl Environ Microbiol 68:1715–1727
108. Kshirsagar HH, Revankar MS, Kamat MY, Lele SS (2002) Cultivation of *Spirulina* in gas induced photobioreactor and isolation of phycobiliproteins. Indian J Biotechnol 1:255–262
109. Weuster-Botz D, Puskeiler R, Kusterer A, Kaufmann K, John GT, Arnold M (2005) Methods and milliliter scale devices for high-throughput bioprocess design. Bioproc Biosyst Eng 28:109–119
110. Puskeiler R, Kaufmann K, Weuster-Botz D (2005) Development, parallelization, and automation of a gas-inducing milliliter-scale bioreactor for high-throughput bioprocess design (HTBD). Biotechnol Bioeng 89:512–523
111. Puskeiler R, Kusterer A, John GT, Weuster-Botz D (2005) Miniature bioreactors for automated high-throughput bioprocess design (HTBD): reproducibility of parallel fed-batch cultivations with *Escherichia coli*. Biotechnol Appl Biochem 42:227–235
112. Knorr B, Schlieker H, Hohmann H-P, Weuster-Botz D (2007) Scale-down and parallel operation of the riboflavin production process with *Bacillus subtilis*. Biochem Eng J 33:263–274
113. Puskeilsr R, Kusterer A, John GT, Weuster-Botz D (2005) Miniature bioreactors for automated high-throughput bioprocess design (HTBD): reproducibility of parallel fed-batch cultivations with *Escherichia coli*. Biotechnol Appl Biochem 42:227–235
114. Vester A, Hans M, Hohmann H-P, Weuster-Botz D (2009) Discrimination of riboflavin producing *Bacillus subtilis* strains based on their fed-batch process performances on a millilitre scale. Appl Microbiol Biotechnol 84:71–76
115. Hoefel T, Wittmann E, Reinecke L, Weuster-Botz D (2010) Reaction engineering studies for the production of 2-hydroxyisobutyric acid with recombinant Cupriavidus necator H 16. Appl Microbiol Biotechnol 88:477–484
116. Kamen AA, Tom RL, Caron AW, Chavarie C, Massie B, Archambault J (1991) Culture of insect cells in helical ribbon impeller bioreactor. Biotechnol Bioeng 38:619–628
117. Jolicoeur M, Chavarie C, Carreau PJ, Archambault J (1992) Development of a helical-ribbon impeller bioreactor for high-density plant cell suspension culture. Biotechnol Bioeng 39:511–521

Adv Biochem Eng Biotechnol (2016) 152: 35–51
DOI 10.1007/10_2014_274

Published Online: 15 July 2014

Novel Hydrogen Bioreactor and Detection Apparatus

Joseph A. Rollin, Xinhao Ye, Julia Martin del Campo, Michael W. W. Adams and Y.-H. Percival Zhang

Abstract In vitro hydrogen generation represents a clear opportunity for novel bioreactor and system design. Hydrogen, already a globally important commodity chemical, has the potential to become the dominant transportation fuel of the future. Technologies such as in vitro synthetic pathway biotransformation (SyPaB)—the use of more than 10 purified enzymes to catalyze unnatural catabolic pathways—enable the storage of hydrogen in the form of carbohydrates. Biohydrogen production from local carbohydrate resources offers a solution to the most pressing challenges to vehicular and bioenergy uses: small-size distributed production, minimization of CO_2 emissions, and potential low cost, driven by high yield and volumetric productivity. In this study, we introduce a novel bioreactor that provides the oxygen-free gas phase necessary for enzymatic hydrogen generation while regulating temperature and reactor volume. A variety of techniques are currently used for laboratory detection of biohydrogen, but the most information is provided by a continuous low-cost hydrogen sensor. Most such systems currently use electrolysis for calibration; here an alternative method, flow calibration, is introduced. This system is further demonstrated here with the conversion of glucose to hydrogen at a high rate, and the production of hydrogen from glucose 6-phosphate at a greatly increased reaction rate, 157 mmol/L/h at 60 °C.

J. A. Rollin · X. Ye · J. M. del Campo · Y.-H. P. Zhang (✉)
Biological Systems Engineering Department, Virginia Tech, 304 Seitz Hall, Blacksburg, VA 24061, USA
e-mail: ypzhang@vt.edu

J. A. Rollin · Y.-H. P. Zhang
Cell Free Bioinnovations Inc, Blacksburg, VA 24060, USA

J. A. Rollin · Y.-H. P. Zhang
Gate Fuels Inc, Blacksburg, VA 24060, USA

M. W. W. Adams
Department of Biochemistry and Molecular Biology, University of Georgia, Athens, GA 30602, USA

Keywords Biohydrogen · In vitro synthetic biosystem for biomanufacturing · Synthetic pathway biotransformation · Hydrogen bioreactor · Continuous hydrogen detection

Contents

1 Introduction 36
2 Setup of Hydrogen Production Bioreactor and Detection Apparatus 38
3 Experimental 41
3.1 Chemicals and Strains 41
3.2 Recombinant Protein Expression and Purification 41
3.3 Enzyme Activity Assays 43
3.4 Preparation of the Enzyme Cocktail 43
3.5 Other Assays 44
4 Hydrogen Production from Glucose via a Synthetic Enzymatic Pathway 44
5 High-Speed Hydrogen Production from Glucose 6-Phosphate 47
6 Large-Scale Hydrogen Production Bioreactor 47
7 Conclusions 49
References 49

1 Introduction

Hydrogen is currently an important commodity chemical. With a global annual market volume of approximately 45 million metric tons per year and a steady increase in demand over the last three decades [1], hydrogen is used in oil refining, fertilizer (ammonia) production, food production, metal treatment, as rocket engine fuel, as electric generator coolant, and for many other industrial purposes [2]. In addition, hydrogen is a target fuel: it is the energy carrier of the hydrogen economy, a vision that seeks to increase transportation energy efficiency greatly, mitigate non-point source pollution, and minimize greenhouse gas emissions [3, 4].

Because the current market for hydrogen primarily serves the petroleum refining and chemical processing industries, 85 % of hydrogen is used near its production site [5]. For such captive markets, large-scale production by steam–methane reforming (SMR) is by far the most economical method. Accordingly, SMR, a process that generates fossil CO_2 emissions as well as carbon monoxide contaminants, currently accounts for approximately 90 % of hydrogen production [6]. For future use as a fuel, important goals include high purity, distributed hydrogen generation, and maximal use of CO_2-free technologies [3]. Purity is especially important, as proton exchange membrane (PEM) fuel cells are poisoned by carbon monoxide. Biohydrogen made from carbohydrates has the potential to meet these requirements.

Biohydrogen is available from several biological routes. Microbial production options include dark fermentation by a pure microorganism and consortia (combining strains specializing in photosynthesis and fermentation), biophotolysis of water, and photodegradation of organic compounds [7]. Systems incorporating photosynthesis are typically limited by the low energy concentration of solar energy and coproduction of oxygen, which inhibits hydrogenase enzymes. Both of these factors result in low volumetric hydrogen productivity [6]. An advanced, porous glass-immobilized cell reactor achieved rates an order of magnitude higher than other studies [8], but scaling and process economics remain a challenge for such designs. Biohydrogen produced via dark fermentation is more promising, but the maximum theoretical yield is 4 mol H_2/mol hexose (the Thauer limit), making cost effectiveness difficult due to the substrate cost when using media containing carbohydrates [9]. Wastewater treatment solves the substrate cost issue, but introduces other challenges, such as coproduction of other products, such as methane and organic acids.

Microbial hydrogen bioreactors are multiphase by nature, incorporating solid-phase reactants, liquid-phase biocatalysts, and gas-phase products. Continuous operation is usually chosen for applications larger than lab scale, often by using continuously stirred tank reactors (CSTRs), although also packed bed, fluidized bed, or membrane reactors have been used in order to reduce hydraulic residence time, a strategy used to prevent the growth of methanogenic organisms [10, Oh, 2004 #336]. Other challenges for large-scale microbial hydrogen production include oxygen inhibition, mixing, and product inhibition. Oxygen inhibition is solved differently by different organisms: some are able to compartmentalize their hydrogenase enzymes, whereas others are strict anaerobes. In either case, minimal oxygen is desired in the fermentation broth. With typical CSTR mixing, substrate addition results in a highly dynamic environment for the cells, with a wide range of concentration throughout the fermentation broth [11]. Hydrogen and CO_2 are gases, which will bubble out of the fermentation above a certain concentration. However, even low levels of dissolved hydrogen can cause problems for most biohydrogen production systems. Solutions to this problem include hydrogen absorption, stripping by elevated reaction temperature, stripping with a gas, and stripping by evaporation or boiling [12], as well as selective permeable membranes. Each of these options increases the cost of the microbial hydrogen-producing bioreactor.

A solution to some of the complexities of microorganisms is to use instead enzymes to catalyze the reaction from carbohydrate substrates directly. This concept was first pioneered by Woodward [13, 14], who demonstrated the conversion of glucose and glucose 6-phosphate to hydrogen using first two, then eleven enzymes (respectively), and Zhang [15], who enabled the use of a lower-cost substrate, starch. This later discovery, of enzymatic oxidation of a nonphosphorylated substrate completely to hydrogen and CO_2, opened up the possibility of economical large-scale enzymatic hydrogen production. Referred to as a synthetic pathway biotransformation (SyPaB), this technology combines purified enzymes from a variety of different hosts (including animal, plant, bacterial, and archaeal)

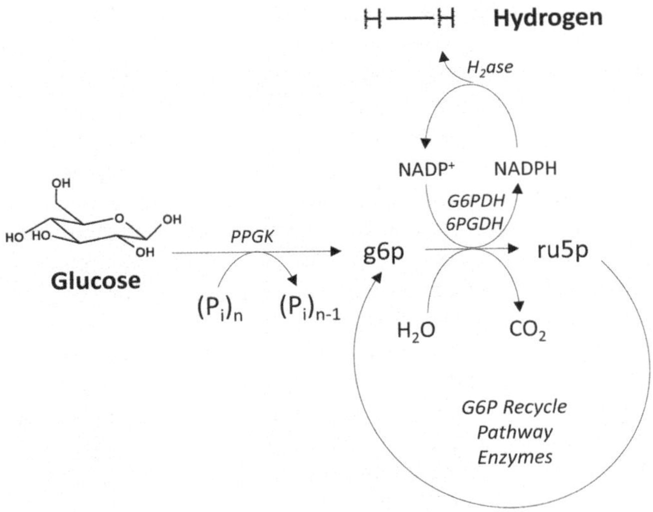

Fig. 1 Synthetic pathway biotransformation (SyPaB) depicting the pathway for conversion of glucose to hydrogen and CO_2. See Rollin et al. [39] for pathways depicting the utilization of additional substrates

to catalyze the high yield of various starting substrates including starch, cello-oligomers, sucrose, and xylose to hydrogen and carbon dioxide. A representative pathway is shown in Fig. 1. This pathway enables production of a 12 mol H_2/mol glucose unit, the theoretical maximum for this molecular conversion and threefold the microbial Thauer limit. Practical yields of over 93 % theoretical have been demonstrated for multiple substrates [16–18]. In this book chapter, the technology features a demonstrated maximum rate of 157 mmol/h/L for the first time, on par with the best biohydrogen rates reported [8].

2 Setup of Hydrogen Production Bioreactor and Detection Apparatus

A system was constructed to study the SyPaB reaction with the following goals: (i) ensure an oxygen-free gas phase in the bioreactor, (ii) remove produced hydrogen with a carrier gas, (iii) ensure constant volume in the bioreactor for long-term operation (preventing evaporative loss or gain of condensation), and (iv) accurately detect a range of hydrogen concentrations, from 1–1,000 ppm, in realtime. A schematic of the system produced is included as Fig. 2a. Our apparatus mimics a similar system at Oak Ridge National Laboratory, which has been described previously in a number of biohydrogen experiments [13, 14, 16, 19–21], with

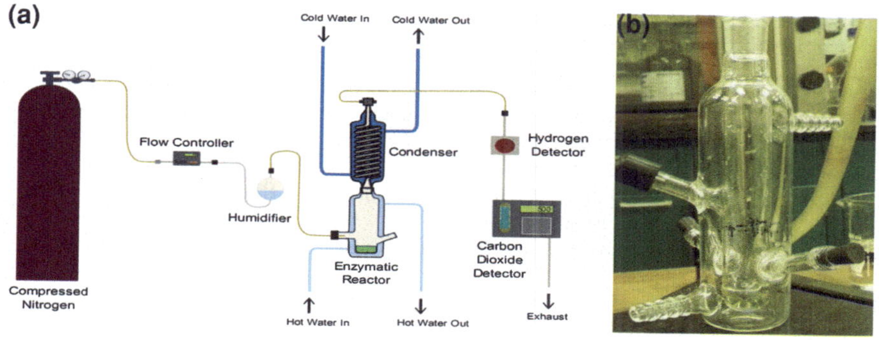

Fig. 2 First-generation, lab-scale hydrogen bioreactor and related hydrogen detection apparatus (**a**). Nitrogen carrier gas is used to prevent oxygen inhibition of hydrogenase. The carrier gas is first humidified before passing through the reactor. A condenser compensates for evaporation of reactor liquid. Hydrogen and carbon dioxide sensors are then used to detect the concentration of these components. The enzymatic reactor (**b**) is a custom glass reactor, which allows temperature control via water bath, carrier gas input, and access to the liquid phase by pipette

some important changes. Specific equipment and software used are summarized in Table 1.

The custom bioreactor designed for this system is shown in Fig. 2b. The reactor is heated with water, warmed in a bath, and pumped through an outer jacket. Actual reactor temperature was found to be 0.4 °C less than the water bath set-point. Gas enters through an upper side port and leaves through the top, where the reactor adjoins a condenser. Lower side injection ports allow addition and removal of the liquid reaction solution, and a stir bar provides sufficient agitation for efficient mass transfer in the system.

In this system, ultra-high–purity (UHP) nitrogen gas is controlled at a rate of 30 mL/min with a gas flow controller (GFC; SKUW-188427, Aalborg, Orangeburg, NY). This carrier gas is first humidified by bubbling through deionized water, after which it enters the reactor. The enzymatic bioreactor is held at a desired temperature (e.g., 50 or 60 °C) with a water jacket and controlled with a water bath (Neslab RTE-7D1, Thermo Scientific, Waltham, MA). Carrier gas plus generated hydrogen and CO_2 next pass through a condenser, held at 21 °C by a second water bath. Condenser temperature was empirically determined to ensure a constant reactor volume. Humidifier, reaction vessel, and condenser were custom fabricated out of borosilicate by the Virginia Tech Department of Chemistry Glass Shop.

Hydrogen detection was conducted using the least costly sensor available, a tin–oxide thermal conductivity detector (TCD) purchased from Figaro (TGS 821, Figaro USA Inc., Glenview, IL). This detector was used as an alternative to simple endpoint gas composition determination by gas chromatography, as is sometimes used in some biohydrogen studies [22, 23]. Such detection methods enable accurate measurement of produced gas, but have the limitations of cost, maintenance, and requiring several minutes per sample, preventing the precise measurement of hydrogen production in systems of quickly changing hydrogen

Table 1 Key hydrogen production and detection apparatus components

Key part	Function	Manufacturer (product name)
Bioreactor	Provide easy access for liquid and solid additions, allow sampling without oxygen introduction, nitrogen-flushed headspace for anaerobic conditions	Virginia Tech department of chemistry (custom glassware)
H_2 detector	Specifically detect hydrogen at concentrations ranging from 1–1000 ppm	Figaro (TGS 821)
Gas flow controller	Ensure steady nitrogen flow rate through system	Aalborg (GFC /SKUW-188427)
Refrigerated water circulator bath	Hold reactor and condenser temperatures steady	Thermo Scientific (Neslab RTE-7D1)
Interface	Convert analog sensor output to digital	National Instruments (USB-6210 DAQ)
LabView software	Continuously log data throughout hydrogen production and calibration experiments	National Instruments (SignalExpress)

production rate. Thermal conductivity is measured in a TCD by comparing the voltage being transmitted across a reference semiconductor with a semiconductor in contact with a sample gas stream. A property of hydrogen gas is extremely high thermal conductivity; this property is used to detect the concentration of hydrogen in the sample stream. A custom aluminum housing for the TCD was manufactured by the Virginia Tech Department of Chemistry. TCD output was routed to a computer for data acquisition using a USB-6210 DAQ (National Instruments, Austin, TX).

As opposed to the ORNL hydrogen detection system, where extensive humidity controls were employed prior to the tin–oxide hydrogen sensor, we found these components to be unnecessary. The key considerations for accurately using a TCD are constant temperature, humidity, and pressure [24]. Constant humidity was maintained by refilling the humidifier before each experiment, and detector temperature was held constant by placing the TCD in an oven held at 44–46 °C. Pressure in the hydrogen sensor is important due to the effect it has on humidity, temperature, and local hydrogen concentration. Here pressure was maintained at slightly below atmospheric by use of a second flow controller at the exhaust.

In similar systems others have employed electrolysis as a means of calibrating the hydrogen detector [14–16, 19, 20]. In this calibration method, a precise current is applied to a Hoffman apparatus or similar electrolysis device, generating H_2 and O_2 from 1 mM KOH. The level of hydrogen detected by TCD may then be calculated by relating the voltage displayed [19]. This method provided consistent results in previous studies, but the accuracy of the electrolysis calculation depends on the accuracy of the assumed faradaic efficiency (not taking into effect inefficiencies caused by the generation of heat and side reactions, including hydrogen peroxide generation). Because this efficiency is not easily measured, and varies between electrolysis systems, we used a different method: flow calibration.

The flow calibration used serial dilution of a UHP hydrogen stream with UHP nitrogen. A flow diagram of this procedure is shown in Fig. 3. In this example, 2.0 mL/min H_2 is mixed with 200 mL/min N_2 to generate a hydrogen mass fraction of 0.01. A rate of 1 mL/min of this mixture is then again diluted, by the addition of 30 mL/min N_2, resulting in a final hydrogen concentration of 330 ppm. Each of these flow rates was controlled by GFC. Using this method, concentrations from below 10 ppm to over 1 % hydrogen were introduced to the TCD and the output voltage recorded (Fig. 4a). This procedure was repeated, and the results were used to build a calibration curve (Fig. 4b). The flow calibration was repeated bimonthly while experiments were being run. In future experiments where higher than 1 % hydrogen is to be detected, use of an electrochemical sensor from Hach Ultra Analytics (Loveland, CO) is recommended; membrane model 2995A is capable of detecting up to 100 % H_2.

3 Experimental

3.1 Chemicals and Strains

All chemicals were reagent grade and purchased from Sigma-Aldrich (St. Louis, MO) and Fisher Scientific (Pittsburgh, PA), unless otherwise noted. *E. coli* BL21 Star (DE3) (Invitrogen, Carlsbad, CA) containing a protein expression plasmid was used to produce all recombinant protein (enzymes #1–11). Luria-Bertani (LB) medium was used for *E. coli* cell growth and recombinant protein expression supplemented with 100 μg/mL ampicillin or 50 μg/mL kanamycin. Oligonucleotides were synthesized by Integrated DNA Technologies (Coraville, IA) and Fisher Scientific. Hydrogenase SH1 from *Pyrococcus furiosus* was provided by Dr. Michael Adams [25].

3.2 Recombinant Protein Expression and Purification

For the preparation of recombinant proteins: two hundred milliliters of LB culture containing 50 μg/mL of kanamycin or 100 μg/mL of ampicillin in 1 L Erlenmeyer flasks were incubated with a rotary shaking rate of 250 rpm at 37 °C. When the absorbance (A_{600}) reached around 1.0, recombinant protein expression was induced by adding isopropyl-β-D-thiogalactopyranoside (IPTG) to a final concentration of 0.01–0.1 mM. The culture was then incubated at 37 °C for 4 h or at 18 °C for 20 h, depending on enzyme-specific optimum expression conditions. Cells were then harvested by centrifugation at 4 °C, washed twice with 50 mM Tris-HCl buffer (pH 7.5), and re-suspended in 15 mL of 30 mM Tris-HCl buffer

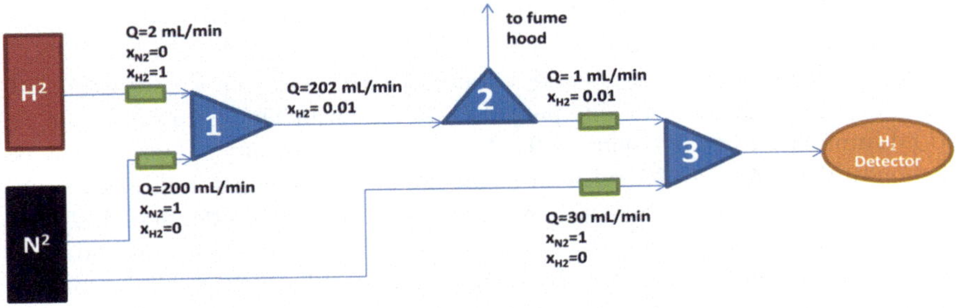

Fig. 3 Flow calibration setup. In this figure green boxes indicate gas flow controllers, and 0.25 in pipe junctions are shown as blue triangles. Pure hydrogen is first mixed with pure nitrogen at junction 1. A subset of this mixture is mixed again with a second stream of pure nitrogen, to yield a total of 30 mL/min gas through the detector. Example flow rates and mass fractions are included above to give a final hydrogen concentration of 330 ppm

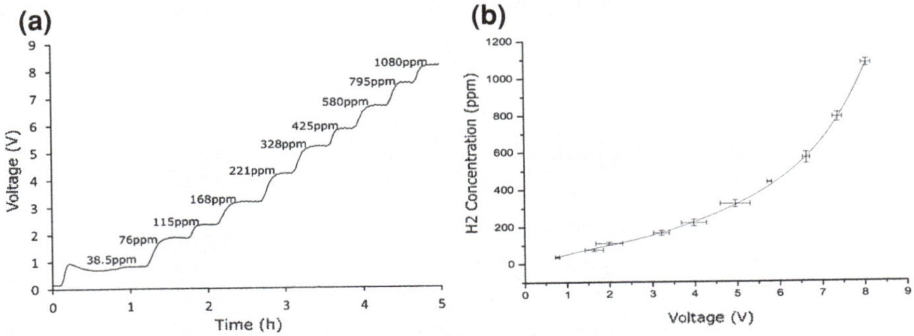

Fig. 4 Example flow calibration (**a**), produced by adjusting the input ratio of pure N_2 and H_2 gases, and calibration curve (**b**), produced from flow calibrations performed in duplicate

(pH 7.5) containing 0.5 M of NaCl and 1 mM of EDTA. Cells were lysed using a Fisher Scientific Sonic Dismembrator Model 500 (5 s pulse on and off, total 360 s, 20 % amplitude) in an ice bath. After centrifugation, the target proteins were purified using several methods, including as His-tag purification, cellulose binding module (CBM) purification followed by intein self-cleavage or ethylene glycol elution, and heat treatment. His-tagged proteins, including G6PDH, 6PGDH, TK, and TAL, were purified using Ni-charged resins (Bio-Rad, Profinity IMAC Ni-Charged Resin). PGM, FBP, PGI, and PPGK were purified by intein self-cleavage or ethylene glycol elution [26–29]. RPI, RPE, TIM, and ALD were purified by heat precipitation conducted at 80 °C for 20 min [30, 31].

3.3 Enzyme Activity Assays

Activity data below include at least one specific activity for each enzyme at 50 °C and/or 60 °C. In the case where one is not listed, a Q10 temperature coefficient of 2 was used for determination of enzyme loading at an uncharacterized temperature.

Thermobifida fusca CBM-PPGK activity was measured based on the generation of glucose 6-phophate from polyphosphate and glucose in a 50 mM HEPES buffer (pH 7.5) containing 4 mM $MgCl_2$, 5 mM D-glucose, and 1 mM polyphosphate at 50 °C for 5 min [28]. The specific activity of CBM-PPGK was 64.6 U/mg at 50 °C.

Geobacillus stearothermophilus G6PDH activity was measured in 100 mM HEPES buffer (pH 7.5) containing 5 mM $MgCl_2$ and 0.5 mM $MnCl_2$, 2 mM glucose 6-phosphate, and 0.67 mM $NADP^+$. The increase in absorbance at 340 nm was measured in 5 min. The specific activity was 35 U/mg at 50 °C [32].

Morella thermoacetica 6PGDH activity was measured in a 50 mM HEPES buffer (pH 7.5) containing 2 mM 6-phosphogluconate, 1 mM $NADP^+$, 5 mM $MgCl_2$, and 0.5 mM $MnCl_2$, at 37 °C for 5 min [32]. The reaction product NADPH was measured at 340 nm. The specific activity was 16 U/mg at 50 °C.

T. maritima RPI activity was assayed by a modified Dische's cysteine–carbazole method. The specific activity was 300 U/mg at 50 °C [30].

T. maritima RPE activity was determined on a substrate D-ribulose 5-phosphate as described previously [33]. The specific activity of RPE was 1.42 U/mg at 50 °C.

T. thermophilus TK activity assay was measured on the substrates of D-xylulose 5-phosphate and D-ribose 5-phosphate. The reactions were carried out in a 50 mM Tris/HCl pH 7.5 buffer containing 0.8 mM D-xylulose 5-phosphate, 0.8 mM D-ribose 5-phosphate, 15 mM $MgCl_2$, 0.03 mM Thiamine pyrophosphate, 0.14 mM NADH, 60 U/mL of TIM, and 20 U/mL of glycerol 3-phosphate dehydrogenase [33]. The specific activity of TK was 1.3 U/mg at 25 °C, or approximately 3.9 U/mg at 50 °C.

T. maritima TAL activity assay was carried out as reported previously. This enzyme had a specific activity of 42 U/mg at 80 °C [34], or 5.25 U/mg at 50 °C.

T. thermophilus TIM activity was measured in 100 mM HEPES pH 7.5 containing 10 mM $MgCl_2$, 0.5 mM $MnCl_2$ at 60 °C for 5 min containing 2 mM D-glyceraldehyde 3-phosphate [35]. The reaction was stopped with $HClO_4$ and neutralized with KOH. The product dihydroxyacetone phosphate was measured by using glycerol 3-phosphate dehydrogenase in the presence of 0.15 mM NADH at 25 °C [35]. The specific activity at these conditions was 870 U/mg at 60 °C.

T. thermophilus ALD was assayed in 100 mM HEPES pH 7.5 containing 10 mM-$MgCl_2$, and 0.5 mM $MnCl_2$ at 60 °C for 5 min with 2 mM of D-glyceraldehyde 3-phosphate in the presence of TIM, FBP, and PGI. The reaction was stopped with $HClO_4$ and neutralized with KOH [35]. The product glucose 6-phosphate was analyzed at 37 °C with a liquid glucose reagent set (Pointe scientific). The specific activity of ALD was 36 U/mg at 60 °C.

T. maritima FBP activity was determined based on the release of phosphate and its specific activity of FBP at 60 °C was 12 U/mg [29].

C. thermocellum PGI activity was assayed at 60 °C in 100 mM HEPES (pH 7.5) containing 10 mM $MgCl_2$ and 0.5 mM $MnCl_2$ with 5 mM fructose 6-phosphate as substrate [27]. After 3 min the reaction was stopped with $HClO_4$ and neutralized with KOH. The product glucose 6-phosphate was analyzed with a liquid glucose hexokinase kit (Pointe scientific). The specific activity of PGI at 60 °C was 1,900 U/mg.

P. furiosus SHI hydrogenase was produced as reported previously [25]. NADPH-based specific activity was 0.8 U/mg at 50 °C.

3.4 Preparation of the Enzyme Cocktail

The reaction buffer contained 100 mM HEPES (pH 7.5), 0.5 mM thiamine pyrophosphate, 10 mM $MgCl_2$, and 0.5 mM $MnCl_2$. For the glucose demonstration experiment, the concentrations of glucose, $NADP^+$, and polyphosphate ($(P_i)_6$, sodium hexametaphosphate) were 2, 4 and 4 mM, respectively. For the high-rate glucose 6-phosphate utilization experiment, the concentrations of glucose 6-phosphate and $NADP^+$ were 100 and 8 mM, respectively. Enzymes were added according to the loadings in Table 2.

To prevent microbial growth, 50 μg/mL of kanamycin was added. Once all other reaction components were combined, substrate was added to start the reaction. The reactor was sealed and magnetic agitation was started, along with nitrogen carrier gas at a flow rate of 30 mL/min. Data acquisition occurred continuously with a TCD. When the oxygen inside the reactor was completely evacuated, the hydrogenase enzymes became uninhibited. Once a sufficient amount of $NADP^+$ was converted to NADPH by the SyPaB pathway, hydrogen production started. During the experiment temperature, carrier gas flow, and hydrogen signal were monitored.

3.5 Other Assays

Soluble protein concentration was measured by the Bio-Rad modified Bradford protein kit with bovine serum albumin as the standard protein. Enzyme molecular weights were verified using 12–15 % SDS-PAGE performed in *Tris*–glycine buffer, as described elsewhere.

4 Hydrogen Production from Glucose via a Synthetic Enzymatic Pathway

To demonstrate the effectiveness of the custom apparatus, hydrogen was generated from glucose. As shown in Fig. 1, in this reaction polyphosphate was used as a phosphate moiety donor, a reaction catalyzed by polyphosphate glucokinase from

Table 2 Enzymes used in hydrogen generation experiments

No.	Enzymes	EC number	Source	Glc loading (U/mL)	G6P loading (U/mL)	References
1	Polyphosphate glucokinase (PPGK)	2.7.1.63	Tfu1811	1	0	[28]
2	Glucose 6-phosphate dehydrogenase (G6PDH)	1.1.1.49	GsG6PDH	1	16	[44]
3	6-phosphogluconate dehydrogenase (6PGDH)	1.1.1.44	Moth1283	1	16	[33]
4	Ribose 5-phosphate isomerase (RPI)	5.3.1.6	Tm1080	1	1.6	[30]
5	Ribulose 5-phosphate 3-epimerase (RPE)	5.1.3.1	Tm1718	1	1.6	[17]
6	Transketolase (TK)	2.2.1.1	Ttc1896	1	1.6	Unpublished data
7	Transaldolase (TAL)	2.2.1.2	Tm0295	1	1.6	[34]
8	Triose phosphate isomerase (TIM)	5.3.1.2	Ttc0581	1	1.6	[33]
9	Fructose bisphosphate aldolase (ALD)	4.1.2.13	Ttc1414	1	1.6	[18]
10	Fructose bisphosphatase (FBP)	3.1.3.11	Tm1415	1	1.6	[29]
11	Phosphoglucose isomerase (PGI)	5.3.1.9	Cthe0217	1	1.6	[27]
12	Hydrogen dehydrogenase (H_2ase)	1.12.1.3	P. furiosus	1	5.6	[25]

The first enzyme loading column is for the proof of concept conversion of glucose, and the second is high-rate production from glucose-6-phosphate

T. fusca (Liao et al. [28]). This kinase is able to use polyphosphate instead of ATP. The product of this reaction, glucose 6-phosphate (g6p), is then oxidized to ribulose 5-phosphate (ru5p) by the action of two dehydrogenases, glucose 6-phosphate dehydrogenase (G6PDH) and 6-phosphogluconolactone (6PGDH), along with the coreduction of $NADP^+$ to NADPH. Pentose phosphate pathway enzymes then convert 6 mol equivalent of ru5p to 5 mol equivalent of g6p. A hydrogenase enzyme completes the cascade, converting NADPH plus free protons to H_2 and recycled $NADP^+$. Equation (1) shows the overall formula for hydrogen production from glucose.

$$C_6H_{12}O_6 + 7H_2O + (HPO_3)_n \rightarrow 12H_2 + 6CO_2 + (HPO_3)_{n-1} + H_3PO_4 \quad (1)$$

The reaction was carried out in 100 mM HEPES buffer (pH 7.5), containing 2 mM glucose, 2 mM polyphosphate, 4 mM $NADP^+$, 10 mM $MgCl_2$, 0.5 mM $MnCl_2$, and 0.5 mM thiamine pyrophosphate. Enzyme loadings were 10 U G6PDH and 6PGDH, and 1 U of all other enzymes. Reactor volume was 1 mL.

The maximum hydrogen generation rate for this reaction was 8.4 mmol H_2 per liter of reactor volume per hour (mmol/L/h) at 50 °C (Fig. 5a). This is over twice the rate observed for previous SyPaB studies conducted at the same substrate loading (2 mM) and similar enzyme loadings. These cases produced maximum

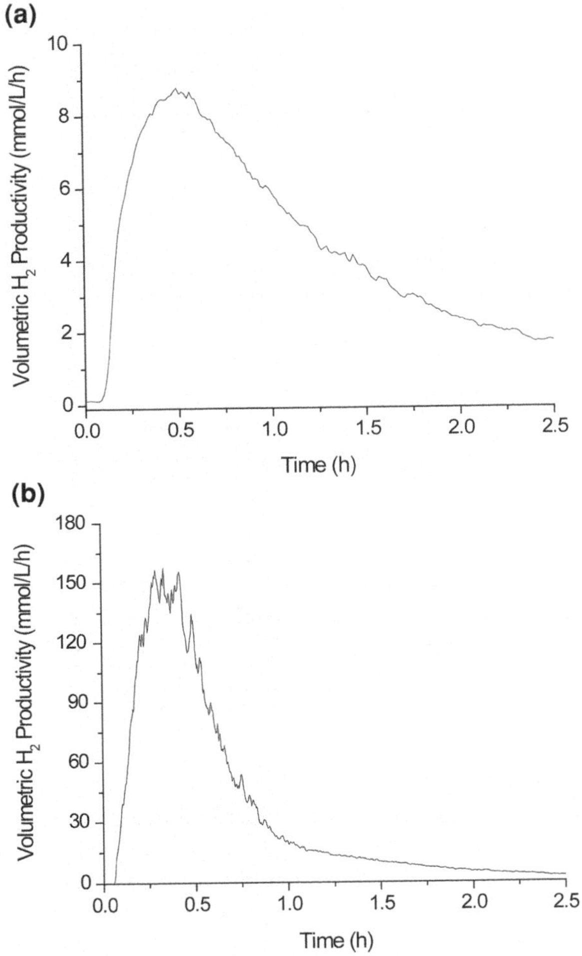

Fig. 5 Production of hydrogen from glucose for the first time (**a**) and high-rate production of hydrogen from glucose-6-phosphate (**b**). The glucose reaction was carried out at is reaction was carried out at 50 °C, with a substrate loading of 2 mM, 4 mM $NADP^+$, and 1 U/mL enzyme loading. The glucose 6-phosphate reaction was conducted at 60 °C, with a substrate loading of 100 mM, 8 mM $NADP^+$, and a high total enzyme loading (Table 2)

rates of 0.44 mmol/L/h from starch at 30 °C [15], 0.72 mmol/L/h from g6p at 30 °C [15], 0.48 mmol/L/h from cellobiose and 3.92 mmol/L/h from cellopentose at 32 °C [16], and 2.2 mmol/L/h from xylose at 50 °C [17].

Another trend apparent among these results is the shape of the curve. Studies with an irreversible phosphorylation reaction (glucose and g6p) as a substrate typically have a sharp peak of initial production, followed by low-level H_2 production for several hours, such as this study and those by Myung et al. [18] and Zhang et al. [15]. Polymeric substrates with a reversible step included in their phosphorylation reaction (i.e., starch) or a long substrate phosphorylation (e.g., xylose) tend to have lower initial peaks, with a period of intermediate H_2 production [15–17]. Production from sucrose, in which half the hexose substrate contains a reversible step, and half of which does not, displays a hybrid profile [18].

5 High-Speed Hydrogen Production from Glucose 6-Phosphate

The rate of hydrogen generation is a primary factor for determining the possible applications for a given technology [36]. Although the rate of 8.4 mmol/L/h observed during the conversion of glucose at 50 °C was higher than the previous maximum rate of 3.92 mmol/L/h [16], higher temperatures and increased substrate concentration, and more enzyme and cofactor loadings offered the potential for additional rate enhancements.

To further increase hydrogen generation rates, we investigated the effects of elevated temperatures, substrate concentration, and key enzyme loadings on the maximum hydrogen generation rate as predicted previously [36]. Most of the enzymes (#2–12) used in the production of hydrogen from glucose are thermostable at temperatures above 50 °C, however #1, PPGK, degrades rapidly above this temperature. For this reason, glucose 6-phoshate was selected as the starting substrate for the high-rate experiments. When more stable phosphorylases or kinases are found or engineered, we expect similar rates will be available from lower-cost substrates as well.

The case where the highest rate of SyPaB hydrogen production was achieved is shown in Fig. 5b. The reaction temperature was increased to 60 °C, the substrate loading was increased to 100 mM glucose 6-phosphate, and the cofactor concentration was increased to 8 mM. Enzyme loadings were increased as summarized in Table 2, where the total reaction volume was 1 mL. The maximum hydrogen rate was 157 mmol/L/h, one of the highest biohydrogen generation rates reported [37]. This reaction rate can be compared to well-known fermentations using a power output per unit volume basis. Examining the maximum hydrogen production rate on the basis of combustible energy produced, 157 mmol H_2/L/h is equivalent to 12.5 W/L, already approaching the energy productivity of a highly mature bioprocess, ethanol fermentation, which is typically 16–30 W/L [38]. In the past decade, SyPaB hydrogen generation has been increased from 0.21 to 157 mmol of H_2 per liter per hour, an improvement of approximately 750-fold. Note that NADP is not stable at 60 °C, resulting in a hydrogen yield lower than the theoretical yield. Therefore, it is essentially important to replace NADP with low-cost and stable biomimetic cofactors in the future [39].

6 Large-Scale Hydrogen Production Bioreactor

The use of purified enzymes simplifies bioreactor design challenges in several regards. Without cell membranes, a major barrier is eliminated in the transport of substrates and products, enabling faster reaction rates, as well as easier sampling and better engineering control [39]. When thermoenzymes are used, higher temperatures may be used, which enable still faster reaction rates, without the difficulties associated with thermophile fermentation. Thermoenzymes also enable the

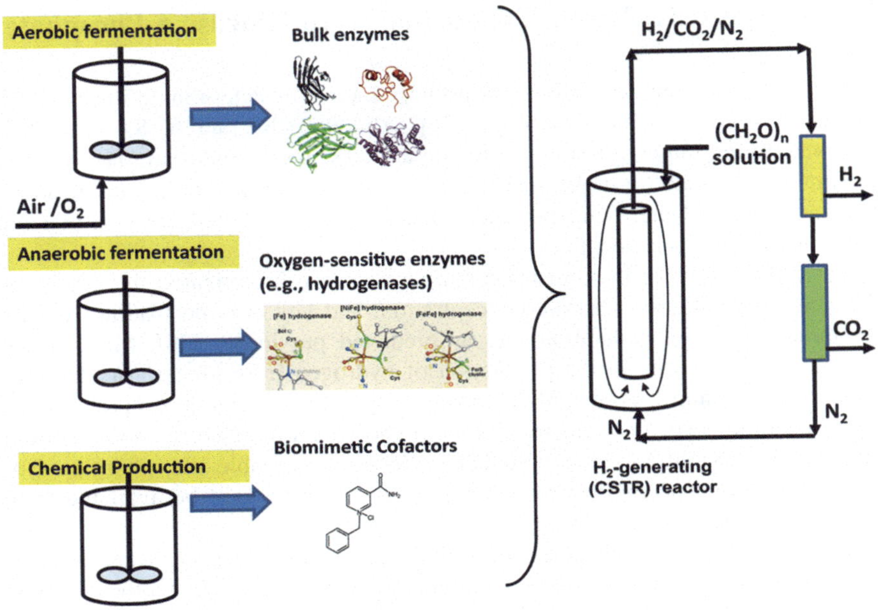

Fig. 6 Reactor systems used for large-scale hydrogen production, including the preparation of enzymes and biomimetic cofactor, continuous hydrogen production in an air-lifted CSTR, and membrane reactors for product separation

possibility of immobilization in a much simpler manner than is possible for microbial systems [40]. Significantly, enzymatic hydrogen production by SyPaB is an endothermic, exergonic reaction [15]. This means heat transfer from nearby operations may be used effectively to upgrade waste heat from air conditioners, refrigerators, and fuel cells to chemical energy H_2, resulting in unmatched thermal conversion efficiency. The challenges that remain for cell-free hydrogen are primarily oxygen and product inhibition, which drive the requirement for an anaerobic reactor with in situ product removal.

In order for biocommodities to be produced economically using in vitro methods, industrial-scale reactors must be employed. Primary concerns at the laboratory scale revolve around elimination of oxygen, temperature control, stable pH, product removal, and easy access, however, factors such as cost and homogeneity become equally important when scaling up. The most common method of minimizing cost and ensuring homogeneity in industrial bioprocesses is to develop continuous methods; a similar alternative is fed-batch, which may be considered a continuous process for periods of time in systems such as enzymatic hydrogen generation, where soluble biocatalysts, cofactors, and reaction buffer are the only components that are not added or removed.

A general proposed flow diagram for fed-batch SyPaB hydrogen generation is presented as Fig. 6. This scheme, although similar to that proposed by Swartz [41], has key differences: in this case the enzymatic hydrogen production occurs in the absence of oxygen, and different bioreactors are used to produce different enzymes instead of one due to different growth conditions. Here there are three biocatalyst sources: anaerobic fermentation for producing oxygen-sensitive enzymes such as hydrogenases; aerobic fermentation for producing most enzymes, such as #1–11; and synthetic chemical production of cofactors. Biomimetic cofactors such as the example molecule shown provide increased thermostability and reduced cost [39].

Selective separation of gaseous products from the carrier gas stream is necessary for continuous operation (Fig. 6) as a pair of membrane separation units; these may also be combined into a single step with the reactor [42], although this may incur additional cost. More complex configurations may also be considered, in order to achieve certain goals. For example, modeling by Ardao and Zeng [43] suggested that SyPaB hydrogen production would benefit from separation of the hydrogenase and other reactions, in order to capitalize on the higher hydrogenase activity at higher temperatures.

7 Conclusions

In summary, a new system for hydrogen generation and detection was constructed, calibrated, and tested. Novel reactor design allowed precise control of the liquid phase, prevented oxygen contamination, and ensured the system was well-mixed and held at a constant temperature, and had a constant volume. A flow calibration system was used to avoid inaccuracies associated with electrolysis-based calibration calculations. Hydrogen generated from glucose represents the first time monomer hexose sugars were utilized by a SyPaB reaction. Finally, engineering principles were used to increase the maximum volumetric hydrogen production rate to 157 mmol/L/h, a rate sufficient for industrial implementation of this technology.

Acknowledgments This work was supported by the Defense University Research Instrumentation Program (DURIP), Shell GameChanger Program, the CALS Biodesign and Bioprocessing Research Center to PZ at Virginia Tech, as well as two NSF STTR I (IIP–1321528) and DOE STTR I (DE-SC0009659TDD) awards to Cell Free Bioinnovations Inc. JR was supported by the Department of Defense through the National Defense Science and Engineering Graduate (NDSEG) Program. MA was supported by the Division of Chemical Sciences, Geosciences and Biosciences, Office of Basic Energy Sciences of the US Department of Energy (grant DE-FG05-95ER20175).

References

1. Lipman T (2011) An overview of hydrogen production and storage systems with renewable hydrogen case studies. Clean Energy States Alliance
2. Ramachandran R, Menon RK (1998) An overview of industrial uses of hydrogen. Int J Hydrogen Energy 23:593–598
3. Ramage MP, Agrawal R (2004) The hydrogen economy: opportunities, costs, barriers and R&D needs. National Research Council of the National Academies, p 15

4. Zhang Y-HP (2013) Next generation biorefineries will solve the food, biofuels, and environmental trilemma in the energy-food-water nexus. Energy Sci Eng 1:27–41
5. Daugherty R (2013) Cost effective distributed hydrogen production systems. VT KnowledgeWorks Strategic Services Division
6. Das D, Veziroğlu TN (2001) Hydrogen production by biological processes: a survey of literature. Int J Hydrogen Energy 26:13–28
7. Hawkes F, Dinsdale R, Hawkes D, Hussy I (2002) Sustainable fermentative hydrogen production: challenges for process optimisation. Int J Hydrogen Energy 27:1339–1347
8. Argun H, Kargi F (2011) Bio-hydrogen production by different operational modes of dark and photo-fermentation: an overview. Int J Hydrogen Energy 36:7443–7459
9. Ducat DC, Sachdeva G, Silver PA (2011) Rewiring hydrogenase-dependent redox circuits in cyanobacteria. Proc Nat Acad Sci U.S.A 108:3941–3946
10. Wang J, Wan W (2009) Factors influencing fermentative hydrogen production: a review. Int J Hydrogen Energy 34:799–811
11. Enfors S-O, Jahic M, Rozkov A, Xu B, Hecker M, Jürgen B, Krüger E, Schweder T, Hamer G, O'beirne D (2001) Physiological responses to mixing in large scale bioreactors. J Biotechnol 85:175–185
12. Van Groenestijn J, Hazewinkel J, Nienoord M, Bussmann P (2002) Energy aspects of biological hydrogen production in high rate bioreactors operated in the thermophilic temperature range. Int J Hydrogen Energy 27:1141–1147
13. Woodward J, Mattingly SM, Danson M, Hough D, Ward N, Adams M (1996) In vitro hydrogen production by glucose dehydrogenase and hydrogenase. Nat Biotechnol 14:872–874
14. Woodward J, Orr M, Cordray K, Greenbaum E (2000) Biotechnology: enzymatic production of biohydrogen. Nature 405:1014–1015
15. Zhang Y-HP, Evans BR, Mielenz JR, Hopkins RC, Adams MWW (2007) High-yield hydrogen production from starch and water by a synthetic enzymatic pathway. PLoS ONE 2:e456
16. Ye X, Wang Y, Hopkins RC, Adams MWW, Evans BR, Mielenz JR, Zhang YHP (2009) Spontaneous high-yield production of hydrogen from cellulosic materials and water catalyzed by enzyme cocktails. ChemSusChem 2:149–152
17. del Martin Campo JS, Rollin J, Myung S, You C, Chandrayan S, Patino R, Adams MWW, Zhang YHP (2013) High-yield production of dihydrogen from xylose by using a synthetic enzyme cascade in a cell-free system. Angew Chem Int Ed 125:4685–4688
18. Myung S, Rollin JA, You C, Sun F, Chandrayan S, Adams MWW, Zhang Y-HP (2014) In vitro metabolic engineering of hydrogen production at theoretical yield from sucrose. Metab Eng Epub:10.1016/j.ymben.2014.1005.1006
19. Iwuchukwu IJ, Vaughn M, Myers N, O'Neill H, Frymier P, Bruce BD (2009) Self-organized photosynthetic nanoparticle for cell-free hydrogen production. Nat Nanotech 5:73–79
20. Millsaps JF, Bruce BD, Lee JW, Greenbaum E (2001) Nanoscale photosynthesis: photocatalytic production of hydrogen by platinized photosystem I reaction centers. Photochem Photobiol 73:630–635
21. Zhang Y-HP (2010) Renewable carbohydrates are a potential high-density hydrogen carrier. Int J Hydrogen Energy 35:10334–10342
22. Krassen H, Schwarze A, Friedrich Br, Ataka K, Lenz O, Heberle J (2009) Photosynthetic hydrogen production by a hybrid complex of photosystem I and [NiFe]-hydrogenase. ACS Nano 3:4055–4061
23. Wu J, Upreti S, Ein-Mozaffari F (2013) Ozone pretreatment of wheat straw for enhanced biohydrogen production. Int J Hydrogen Energy 38(25):10270–10276
24. Figaro (2004) TGS 821—special sensor for hydrogen gas product information. Figaro USA Inc
25. Chandrayan SK, McTernan PM, Hopkins RC, Sun JS, Jenney FE, Adams MWW (2012) Engineering hyperthermophilic archaeon *Pyrococcus furiosus* to overproduce its cytoplasmic NiFe–hydrogenase. J Biol Chem 287:3257–3264

26. Wang Y, Zhang Y-HP (2010) A highly active phosphoglucomutase from *Clostridium thermocellum*: cloning, purification, characterization, and enhanced thermostability. J Appl Microbiol 108:39–46
27. Myung S, Zhang X-Z, Zhang Y-HP (2011) Ultra-stable phosphoglucose isomerase through immobilization of cellulose-binding module-tagged thermophilic enzyme on low-cost high-capacity cellulosic adsorbent. Biotechnol Prog 27:969–975
28. Liao HH, Myung S, Zhang Y-HP (2012) One-step purification and immobilization of thermophilic polyphosphate glucokinase from *Thermobifida fusca* YX: glucose-6-phosphate generation without ATP. Appl Microbiol Biotechnol 93:1109–1117
29. Myung S, Wang Y, Zhang Y-HP (2010) Fructose-1, 6-bisphosphatase from a hyper-thermophilic bacterium *Thermotoga maritima*: characterization, metabolite stability, and its implications. Process Biochem 45:1882–1887
30. Sun FF, Zhang XZ, Myung S, Zhang Y-HP (2012) Thermophilic *Thermotoga maritima* ribose-5-phosphate isomerase RpiB: optimized heat treatment purification and basic characterization. Protein Expr Purif 82:302–307
31. Myung S, Zhang Y-HP (2013) Non-complexed four cascade enzyme mixture: simple purification and synergetic co-stabilization. PLoS ONE 8:e61500
32. Zhu ZG, Sun F, Zhang X, Zhang Y-HP (2012) Deep oxidation of glucose in enzymatic fuel cells through a synthetic enzymatic pathway containing a cascade of two thermostable dehydrogenases. Biosens Bioelectron 36:110–115
33. Wang Y, Huang W, Sathitsuksanoh N, Zhu Z, Zhang Y-HP (2011) Biohydrogenation from biomass sugar mediated by in vitro synthetic enzymatic pathways. Chem Biol 18:372–380
34. Huang SY, Zhang Y-HP, Zhong JJ (2012) A thermostable recombinant transaldolase with high activity over a broad pH range. Appl Microbiol Biotechnol 93:2403–2410
35. You C, Myung S, Zhang Y-HP (2012) Facilitated substrate channeling in a self-assembled trifunctional enzyme complex. Angew Chem Int Ed 51:8787–8790
36. Zhang Y-HP (2011) Simpler is better: high-yield and potential low-cost biofuels production through cell-free synthetic pathway biotransformation (SyPaB). ACS Catal 1:998–1009
37. Rittmann S, Herwig C (2012) A comprehensive and quantitative review of dark fermentative biohydrogen production. Microb Cell Fact 11:115
38. Querol A, Fleet GH (eds) (2006) The yeast handbook: yeasts in food and beverages. Springer, Berlin
39. Rollin JA, Tam W, Zhang YP (2013) New biotechnology paradigm: cell-free biosystems for biomanufacturing. Green Chem 15:1708–1719
40. Myung S, You C, Zhang YP (2013) Recyclable cellulose-containing magnetic nanoparticles: immobilization of cellulose-binding module-tagged proteins and synthetic metabolon featuring substrate channeling. J Mater Chem B 15:1708–1719
41. Swartz JR (2013) SBE supplement: biochemical engineering—cell-free bioprocessing. In: Chemical engineering progress. American Institute for Chemical Engineers, New York, USA
42. Prazeres D, Cabral J (1994) Enzymatic membrane bioreactors and their applications. Enzyme Microb Technol 16:738–750
43. Ardao I, Zeng A-P (2013) In silico evaluation of a complex multi-enzymatic system using one-pot and modular approaches: application to the high-yield production of hydrogen from a synthetic metabolic pathway. Chem Eng Sci 87:183–193
44. Zhu Z, Tam TK, Sun F, You C, Zhang Y-HP (2014) A high-energy-density sugar biobattery based on a synthetic enzymatic pathway. Nat Commun 5:3026

Adv Biochem Eng Biotechnol (2016) 152: 53–74
DOI 10.1007/10_2014_282

Published Online: 6 November 2014

Cell Processing Engineering for Regenerative Medicine

Noninvasive Cell Quality Estimation and Automatic Cell Processing

Mutsumi Takagi

Abstract The cell processing engineering including automatic cell processing and noninvasive cell quality estimation of adherent mammalian cells for regenerative medicine was reviewed. Automatic cell processing necessary for the industrialization of regenerative medicine was introduced. The cell quality such as cell heterogeneity should be noninvasively estimated before transplantation to patient, because cultured cells are usually not homogeneous but heterogeneous and most protocols of regenerative medicine are autologous system. The differentiation level could be estimated by two-dimensional cell morphology analysis using a conventional phase-contrast microscope. The phase-shifting laser microscope (PLM) could determine laser phase shift at all pixel in a view, which is caused by the transmitted laser through cell, and might be more noninvasive and more useful than the atomic force microscope and digital holographic microscope. The noninvasive determination of the laser phase shift of a cell using a PLM was carried out to determine the three-dimensional cell morphology and estimate the cell cycle phase of each adhesive cell and the mean proliferation activity of a cell population. The noninvasive discrimination of cancer cells from normal cells by measuring the phase shift was performed based on the difference in cytoskeleton density. Chemical analysis of the culture supernatant was also useful to estimate the differentiation level of a cell population. A probe beam, an infrared beam, and Raman spectroscopy are useful for diagnosing the viability, apoptosis, and differentiation of each adhesive cell.

Keywords Regenerative medicine · Cell processing · Noninvasive · Quality · Automatic

---

M. Takagi (✉)
Division of Biotechnology and Macromolecular Chemistry,
Graduate School of Engineering, Hokkaido University,
N13W8 Kita-ku, Sapporo 060-8628, Japan
e-mail: takagi-m@eng.hokudai.ac.jp

Contents

1 Introduction: Cell Processing Engineering 54
2 Automatic Cell Processing 55
3 Noninvasive Cell Quality Estimation 57
3.1 Analysis of Two-Dimensional (2D) Cell Morphology 58
3.2 Determination of Three-Dimensional (3D) Cell Morphology 60
3.3 PLM and Its Application 61
3.4 Discrimination of Cancer Cells by PLM 63
3.5 Chemical Analysis of Culture Components 68
4 Conclusion and Perspective 71
References 72

1 Introduction: Cell Processing Engineering

Cultivation engineering was established by many researchers aiming at the industrialization of the production of useful substances such as amino acids, antibiotics, and nucleic acids with the use of microbes. The research included medium design, reactor design, flask design, oxygen transfer theory, sensor design, modeling of microbial reactions, and optimization of microbial processes. Cultivation engineering for animal cells was developed from 1980 in order to industrialize the production of animal cell products such as interferon, interleukins, cytokines, tissue plasminogen activator, and therapeutic antibodies, which included the development of serum-free media, new aeration systems without sparging, and the designs of microcarriers for adhesion-dependent cells, reactors, and impellers design. The widespread use and industrialization of regenerative medicine have been highly expected because of the recent rapid progress in the fields of basic life sciences concerning human cells (Fig. 1). Because the control of not only the proliferation but also differentiation of human cells and their three-dimensional (3D) culture into tissue are necessary for the realization of regenerative medicine in which cells are the final product, some new technologies, namely "cell processing engineering," other than the conventional cultivation engineering of animal cells in which cell metabolites are the final products, should be developed.

Cell processing engineering is the cultivation engineering required for the production of cells and tissue for transplantation such as cell therapy and regenerative medicine. Cell processing engineering may include the material design (cell, scaffolds, medium validity) and cultivation process design (3D culture, coculture and cell sheet technology etc.), and they have been well investigated [1, 2]. Besides them, industrialization technologies including automatic cell processing and cell quality estimation have not been well investigated so far.

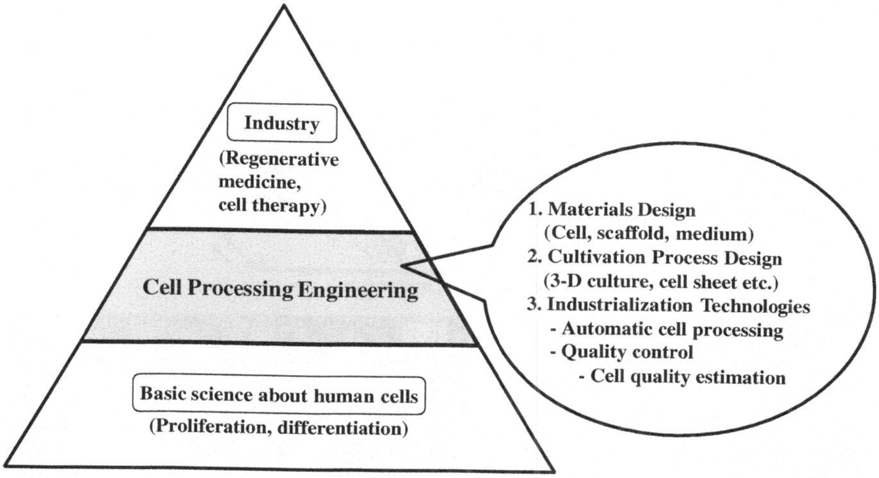

Fig. 1 Subjects of cell processing engineering

2 Automatic Cell Processing

Culture of cells for transplantation without contamination is critical for the use of such cells in regenerative medicine. For this reason, cells are cultured in a cell processing center (CPC) in conformity with the good manufacturing practice (GMP). In the industrial application of cell processing technology, manual cell processing in a CPC, as shown in Fig. 2a, has many problems such as the large initial capital investment, the possibility of human errors in manual processing, the possibility of pathogen transmission from operators to cells and tissues, and the high labor cost for skilled operators. On the other hand, the introduction of automated cell culture systems is expected to reduce the required size of a clean room, lower the degree of cleanliness of the clean room, and markedly reduce the chance of errors and the possibility of pathogen transmission from operators. The main form of regenerative medicine is autologous transplantation in which the patients' own cells are collected, cultured, and then transplanted back into the patients. When there are specimens collected from two different donors, two cell processing rooms, each usually having a clean bench and an incubator, are required to prevent cross-contamination between the specimens.

Automatic cell processing should be established for the industrialization of regenerative medicine, because it may be necessary to skip the manual cell processing in the cell processing center and establish a more stable, safer, and low-cost cell processing. There are three possible types of automated cell culture systems for such a case, as shown in Fig. 3. Type A is designed to automate only the procedures performed by an operator at a clean bench (e.g., medium exchange and subculture). This type of system is considered to have little impact on reducing the labor cost

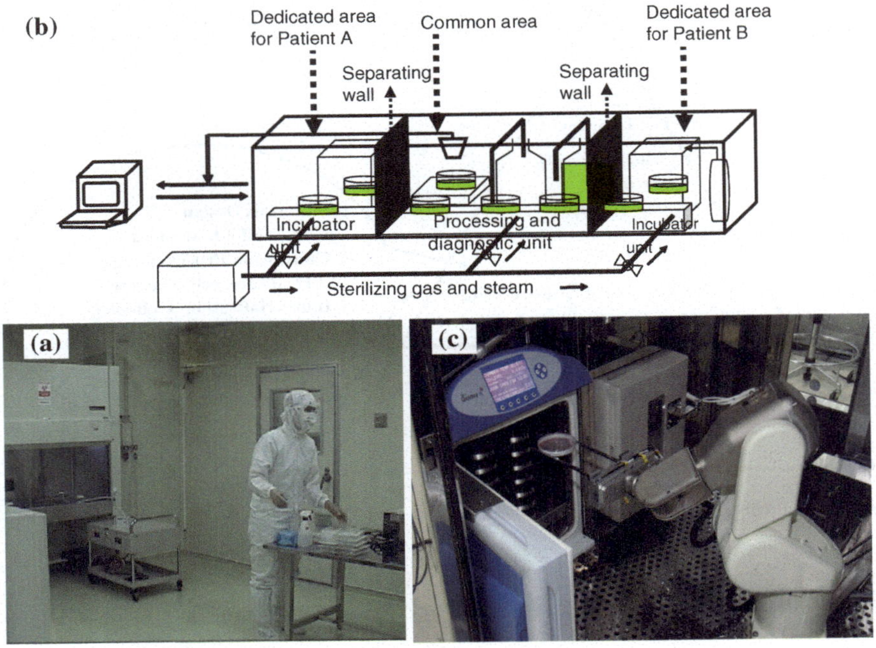

Fig. 2 Image of automated cell culture system used for cells from multiple patients. **a** Manual cell processing in CPC. **b** Schematic of automated cell culture system used for cells from multiple patients. **c** Cell processing procedure performed by clean robot

Fig. 3 Classification of automated cell culture system for transplantation

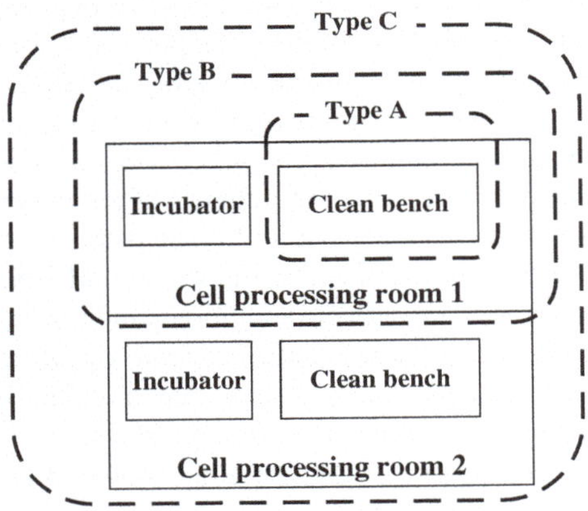

because the extent of automation is limited. Type B is designed to automate the transfer of culture vessels between an incubator and a clean bench and the microscopy observation of cells in addition to the procedures performed by an operator at a clean bench (e.g., medium exchange and subculture). Because of the need for sterilization, specific closed culture vessels are often used in this type of system. Type C is designed to automate the procedures that must be performed in multiple cell processing rooms in addition to the procedures performed by an operator at a clean bench (e.g., medium exchange and subculture), the transfer of culture vessels between an incubator and a clean bench, and the microscopy observation of cells. This type of system is economically efficient because it is designed such that not only the microscope for cell observation and the centrifuge but also the clean bench can be shared for the processing of two or more specimens, leading to reductions in labor cost and capital investment for cell culture.

It may be necessary for an automatic cell processing machine to manipulate cells from plural donors at the same time in order to decrease the machine cost effectively. A Type C automatic cell processing machine that can manipulate cells derived from plural donors without cross-contamination was developed [3]. The example of Type C systems (Fig. 2b), an automated cell culture system for regenerative medicine, has an inner structure that can be separated into multiple incubator units and one processing and diagnostic unit in order to prevent cross-contamination. This system is equipped with a multijoint robot that can accurately reproduce the subtle movements of human operators in manual cell processing procedures (Fig. 2c), as well as a charge-coupled device (CCD) camera to record cell images. A demonstrative test of this type of system is carried out at the CPC of various laboratories, such as the Center for Advanced Medicine, Shinshu University Hospital, and Chulalongkorn University in Thailand [3].

3 Noninvasive Cell Quality Estimation

The system of quality control might also be essential for the industrialization of regenerative medicine. The quality and heterogeneity of cells should be estimated before transplantation, because cultured cells are not homogeneous but heterogeneous (Fig. 4). For example, in the case of cartilage regeneration using bone marrow mesenchymal stem cells (MSCs), not only MSCs but also other kinds of adherent cell such as endothelial cells should be isolated from bone marrow, and MSCs may differentiate into certain lineages even during proliferation culture [4]. Moreover, not all MSCs could differentiate into chondrocytes at the same time [5]. Thus, the quality and heterogeneity of cells should be estimated before transplantation from the viewpoint of quality control.

Viability, proliferation activity, and differentiation level are important parameters concerning cell quality, and there are many methods to estimate cell quality, such as flow cytometry, colony assay, and immune staining. However, all these methods are invasive. Most protocols of regenerative medicine are autologous cell system. A minimum amount of cells are cultivated in the autologous cell system, and we cannot discard the cells for the estimation of cell quality. Thus, the estimation of cell quality should be performed noninvasively. The noninvasive estimation of cell

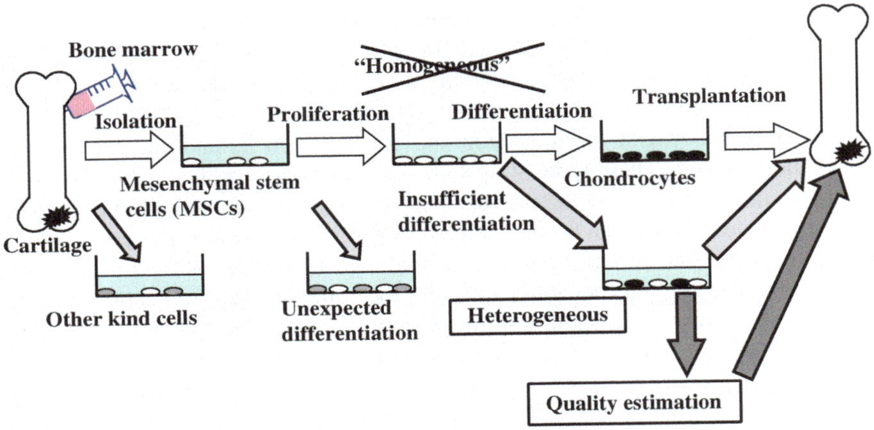

Fig. 4 A problem on quality control in regenerative medicine in the case of cartilage regeneration using mesenchymal stem cells

quality may also help in the process control of adherent cell cultivation and the high-throughput screening of candidates of pharmaceuticals using mammalian cells.

Consequently, the technology for the noninvasive quality estimation of adherent mammalian cells should be developed for the industrialization of regenerative medicine and drug development.

3.1 Analysis of Two-Dimensional (2D) Cell Morphology

There are several reports showing that the function of cells is correlated with their morphology, and images of cell morphology can be obtained noninvasively using a conventional inverted microscope. For example, the disruption of the cytoskeletal structure of chondrocytes by staurosporine has been reported to induce morphological changes and enhance type II collagen expression [6]. The binding of concanavalin A to membrane-bound transferrin-like protein (MTF) on C3H10T1/2 cells, which is a murine cell line of mesenchymal stem cells, was reported to cause a cell-shaped change from fibroblastic to spherical together with increases in the expression levels of aggrecan and type II collagen genes [7].

With the aim of developing a noninvasive method of monitoring the differentiation of MSCs to chondrocytes, the correlation between cell morphology and aggrecan mRNA expression (a marker of chondrocytes) was investigated [5]. Polygonal index, a morphological parameter, was defined as the ratio of cell adhesion area versus the square of the major cell axis, and cells that have an adhesion area larger than 4,000 μm^2 and a polygonal index larger than 0.3 are considered large polygonal cells. Approximately 80 % of the large polygonal cells were negative for MSC marker (CD90, CD166) expression, and the aggrecan mRNA expression level of the large polygonal cells was markedly higher than that of cells with other morphologies. The cells expressing aggrecan mRNA at a high level in the culture in which the differentiation of MSCs to chondrocytes occurred may be characterized using the polygonal index and cell adhesion area. Indeed, cell morphology tended to change from fibroblast like to polygonal and the percentage of large polygonal cells increased almost in proportion to the aggrecan mRNA expression level during the differentiation of MSCs to chondrocytes (Fig. 5), namely the change of cell morphology caused by the differentiation was qualified by the parameter of polygonal index. Thus, the level of differentiation level of MSCs to chondrocytes could be noninvasively estimated on the basis of the percentage of large polygonal cells.

Automated visual tracking of cell populations in vitro using time-lapse phase-contrast microscopy enables the quantitative, systematic, and high-throughput measurements of cell behaviors. These measurements include the spatiotemporal quantification of cell migration, mitosis, and apoptosis. The combination of low signal-to-noise ratio of phase-contrast microscopy images, high and varying

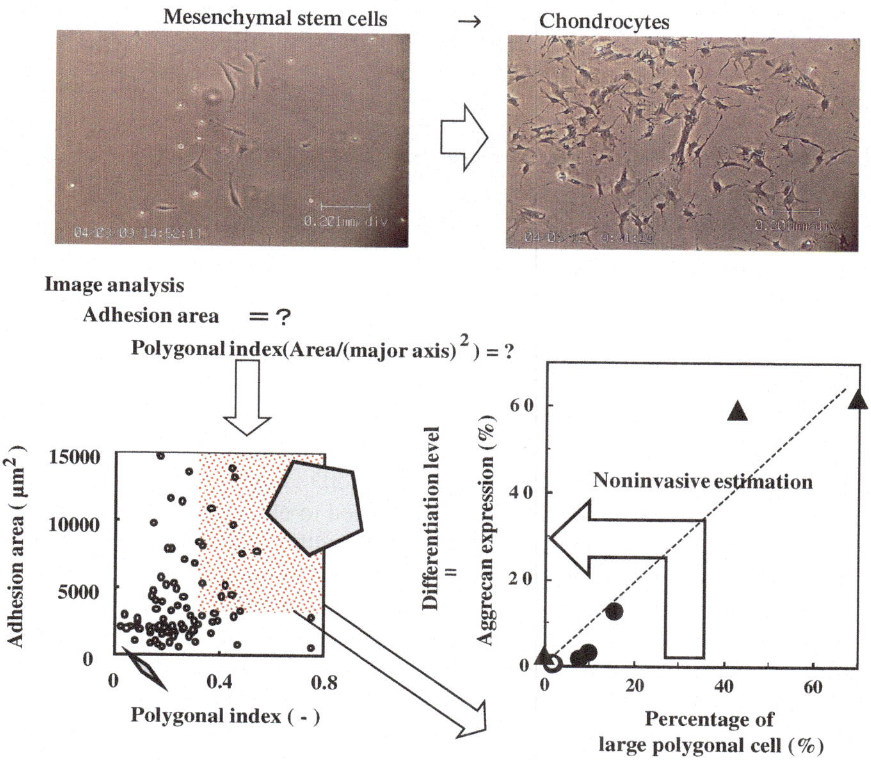

Fig. 5 Noninvasive estimation of differentiation level using two-dimensional cell morphology analysis

densities of cell cultures, topological complexities of cell shapes, and wide range of cell behaviors poses many challenges to existing tracking techniques.

A fully automated multitarget tracking system that can efficiently cope with these challenges, while simultaneously tracking and analyzing thousands of cells observed using time-lapse phase-contrast microscopy, was reported. The system combines bottom-up and top-down image analyses by integrating multiple collaborative modules. The system, which was tested using a variety of cell populations, achieved a tracking accuracy in the range of 86.9–92.5 % [8].

Most of the image analysis has focused on a few parameters concerning cell morphology selected by experts. Bioinformatic analysis to select the best parameter combination for predicting cell quality was shown. By using the fully automatic cell culture and monitoring system (Nikon Corporation), the time-lapse phase-contrast cell images (of more than 4,000 cells) of four different cell lines (fibroblasts, sarcomas, myoblasts, and keratinocytes) as models of cell quality prediction were obtained. By employing the image analysis combined with the bioinformatic parameter (e.g., length of major axis, minor axis, adhesion area, circularity, time-lapse change of these parameters, and so on) selection methodologies, predictions of the contamination percentage of sarcoma cells in fibroblasts, differentiation rate of myoblasts in the future, and contamination rate of fibroblasts in keratinocytes from the early images of cultured cells were performed [9].

3.2 Determination of Three-Dimensional (3D) Cell Morphology

Adherent mammalian cells have not only 2D but also 3D morphology. Analysis of the 3D cell morphology can add useful information to the estimation of cell quality such as cell cycle and canceration as mentioned in later Sects. (3.3.2 and 3.4.1) because cell height is a component of phase shift of a cell. There are several methods for the noninvasive determination of 3D cell morphology.

A method aimed at the quantitative measurement of the optical properties of objects from differential interference contrast (DIC) microscopic images by reconstructing the specimen's optical properties over a 3D volume was reported [10].

An atomic force microscope (AFM) can be used to observe the 3D morphology of adherent animal cells [11]. For example, the differences in 3D morphology between Chinese hamster ovary (CHO) cells cultivated under various osmolarities were determined under an AFM after the fixation of the cells [12]. However, AFM observation is considered invasive for cells, because the 3D observation of adherent animal cells using an AFM requires a long time and fixation treatment.

A digital holographic microscopy method for the integral refractive index determination of living single cells in a cell culture medium was reported. The obtained refractive index information was applied to full field thickness and shape determination of adherent pancreas tumor cells [13]. However, the digital holographic microscopy method suffers from the disadvantages of weakness against vibration and low resolution and accuracy, compared with the phase-shifting laser microscopy (PLM) as discussed below.

3.3 PLM and Its Application

3.3.1 Determination of 3D Cell Morphology by PLM

Recently, PLM has been developed [14]. A biprism, located between the magnifying lens and the observation plane, was used as a beam splitter (Fig. 6). The biprism was laterally translated to introduce phase shifts required for quantitative phase measurement by a phase-shifting technique. Namely, the phase shift (ΔΦ) caused by the difference in refractive indices between the sample and the reference expressed in Eq. (1) can be determined by PLM.

$$\Delta\Phi = 2\pi d \times \frac{n_1 - n_0}{\lambda_0} \quad (1)$$

Here, ΔΦ is the phase shift (−), d is the thickness of the sample (nm), λ_0 is the wavelength of the laser (nm), and n_1 and n_0 are the refractive indices of the sample and reference (−), respectively.

The noninvasive measurement of the 3D morphology of adhered animal cells by PLM, in which the phase shift for each pixel in the view field was determined, was

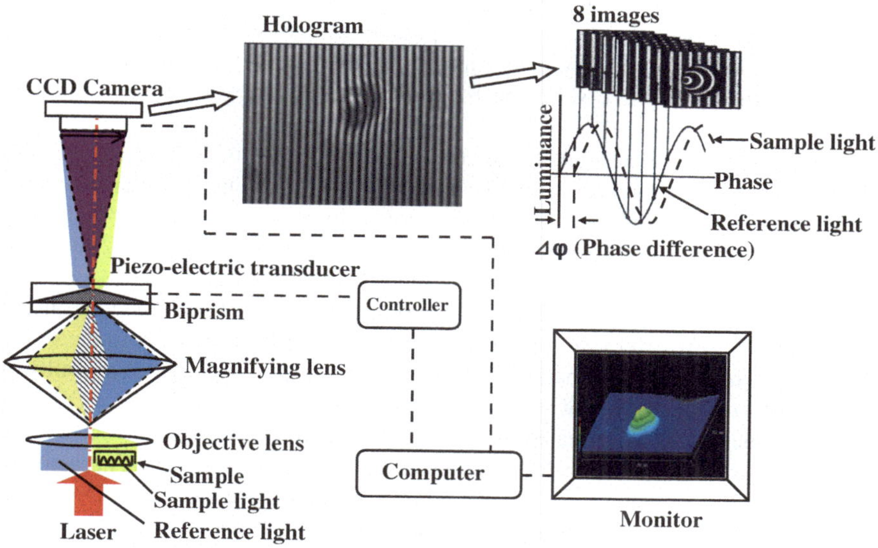

Fig. 6 Schematic diagram of phase-shifting laser microscope

carried out [15]. By employing saline with different refractive indices instead of the culture medium, the refractive index of the cells, which is necessary for the determination of cell height, was determined by PLM. The observed height of CHO cells cultivated under high osmolarity was lower than that of the cells cultivated under physiological osmolarity, which is in agreement with previous data observed under AFM. The maximum heights of human bone marrow mesenchymal stem cells and human umbilical cord vein endothelial cells measured using PLM and AFM agreed well with each other. The maximum height of nonadherent spherical CHO cells observed by PLM was comparable to the cell diameter measured using a phase-contrast inverted microscope. Laser irradiation, which was necessary for the observation by PLM, did not affect the 3D cell morphology. Thus, the 3D morphology of adhered animal cells could be noninvasively determined by PLM.

3.3.2 Estimation of Cell Cycle and Proliferation Activity by PLM

The growth activity of adherent cells is one of the most important cell parameters. However, estimations of growth activity by determining the uptake rate of nucleic acids such as bromodeoxyuridine and cell cycle analysis using a flow cytometer and propidium iodide-stained cells are invasive and time-consuming.

A study using scanning and transmission electron microscopes revealed that the height and roundness of adherent rat kangaroo cells (PtK2) increase in late anaphase [16]. Observation using an AFM [12] and PLM [15] revealed that the growth rate and height of adherent CHO cells cultivated under a high osmolarity of 400 mOsm/L are lower than those of cells cultivated under the physiological osmolarity of 300 mOsm/L. DNA condensation in cells in the G_2/M cell cycle phase was expected to result in a higher cellular refractive index. The laser phase shift of adherent animal cells, which is a product of cell height and cell refractive index, could be noninvasively determined by PLM [14, 15]. These findings suggested that the laser phase shift and roundness of cells depend on the cell cycle phase.

The simultaneous determination of the cell cycle phase of individual adherent CHO cells using a fluorescence microscope after staining with 4', 6-diamidine-2'-phenylindole dihydrochloride and bromodeoxyuridine and the laser phase shift determined by PLM revealed that the laser phase shift of cells in the G2/M phase was markedly higher than that of cells in the G1 and S phases [17].

This was also true for MSCs (Fig. 7a). Even in the cultures of MSCs synchronized to G_0/G_1 and G_2/M cell cycle phases, the laser phase shift of the cells in the G_2/M phase was markedly higher than that of the cells in the G_0/G_1 phase. The results of the analysis of cultures of MSCs from different donors with the addition of FGF2 at different concentrations revealed that there was a marked negative correlation between the average phase shift and mean generation time (Fig. 7b). The actively proliferating cell population might contain higher content of cells in the G_2/M phase, and it may be the reason for the negative correlation. Thus, it was possible to

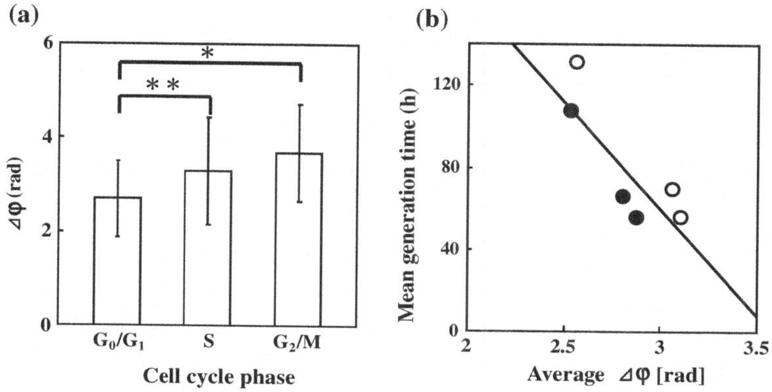

Fig. 7 Analysis of cell cycle phase and proliferation activity of MSCs by phase-shifting laser microscopy. **a** Effect of cell cycle phase on laser phase shift. By employing the threshold values for fluorescence intensities and classifying the cells into their respective cell cycle phase, the laser phase shift ($\Delta\varphi$) in each cell cycle phase was determined (average ± SD, *; $p < 0.01$, **; $p < 0.05$). **b** Relationship between mean generation time and average phase shift. Mean generation times of MSCs from a donor cultivated in the presence of various concentrations of FGF2 (open circles) and MSCs from various donors (closed circles) were plotted against their average phase shift ($\Delta\varphi$). The correlation coefficient was 0.71 [18]

estimate noninvasively the proliferation activity of MSCs by measuring the phase shift by PLM [18].

3.4 Discrimination of Cancer Cells by PLM

3.4.1 Difference in Phase Shift Between Human Normal Cells and Malignant Tumor Cells

Besides the differentiation degree, cell cycle phase, and proliferation activity, it might be important to evaluate whether the cultured cells contain cancer cells. The standard technique to investigate whether the cultured cells contain cancer cells is to determine the structural and numerical aberrations of chromosome using CHO cells and human peripheral blood lymphocytes [19]. However, these methods are destructive and time-consuming. Thus, a noninvasive evaluation method to determine whether cancer cells are contained is necessary.

A study using an AFM revealed that the elasticities of two kinds of normal human bladder epithelial cells (Hu609 and HCV29) were ten times higher than those of three kinds of cancerous ones (Hu456, T24, and BC3726) [20, 21]. The reason for the difference in elasticity may be the difference in the amount of cytoskeleton elements, because the cancerous cells were reported to be softer and contain less amount of cytoskeleton than normal cells. If the difference in the amount of cytoskeleton results in the difference in the refractive index of the cells, cancer cells may be distinguished from normal cells by phase shift determination using PLM.

However, there has been little research on the difference in the phase shift between cancer and normal cells or on their discrimination by phase shift measurement. Moreover, it might be essential to evaluate noninvasively the contamination of cancer cells and determine the percentage of cancer cells in the cell population derived from normal cells, from the viewpoint of quality control of cells cultivated for transplantation.

Consequently, it was investigated to distinguish noninvasively cancerous cells from normal cells in adherent culture by phase shift measurement using PLM and how to determine the percentage of cancer cells in a mixture of cancer and normal cells [22].

Primary normal human prostate epithelial cells (PREC), a human prostatic carcinoma epithelial cell line (PC-3), cryopreserved human hepatocytes (HCM), and four kinds of human hepatocellular carcinoma (Hep3B, PLC [23], HLF [24], and Huh7 [25]) were used. The cells were plated on a dish with grids (11.8 cm^2; SARSTEDT, Nümbrecht, Germany) at a density of 0.15×10^4 cells/cm^2 using the respective medium and incubated at 37 °C under 5 % CO_2 atmosphere for 24 h.

To investigate the difference in the phase shift between PREC and PC-3 cells, the laser phase shifts of these cells were analyzed by PLM. There was no marked difference in 2D morphology between these cells. However, the average phase shifts of PREC and PC-3 cells were 5.05 and 3.10 rad, respectively (Fig. 8a), and the difference showed 1 % significance. That result showed that the phase shift of cancerous PC-3 cells was markedly lower than that of normal PREC.

To investigate the difference in the phase shift between normal HCM cells and cancerous human hepatocyte cell lines (Hep3B, HLF, Huh7, and PLC), the laser phase shifts of these cells were analyzed by PLM. There was no marked difference in 2D morphology between these five kinds of cells. On the other hand, the average phase shifts of the cancer cell lines (Hep3B, HLF, Huh7, and PLC, 3.06, 3.56, 4.09, and 3.72 rad respectively) were markedly smaller than that of normal HCM cells (5.87 rad), and the difference showed 1 % significance (Fig. 8b). That result showed that the phase shifts of the cancer hepatocyte cells were markedly smaller than that of normal HCM cells.

Those two results indicate that adherent cancer cells and normal cells could be noninvasively discriminated by phase shift measurement using PLM, even if their 2D morphologies were similar.

One of the reasons for the difference in phase shift between adherent cancer cells and normal cells may be the difference in the amount of cytoskeleton elements between them. Indeed, decreasing the actin content of human pancreatic ductal adenocarcinoma cell lines by adding latrunculin B, which disrupts the actin cytoskeleton, was reported to decrease the phase shifts of the cells as determined by digital holographic microscopy [13].

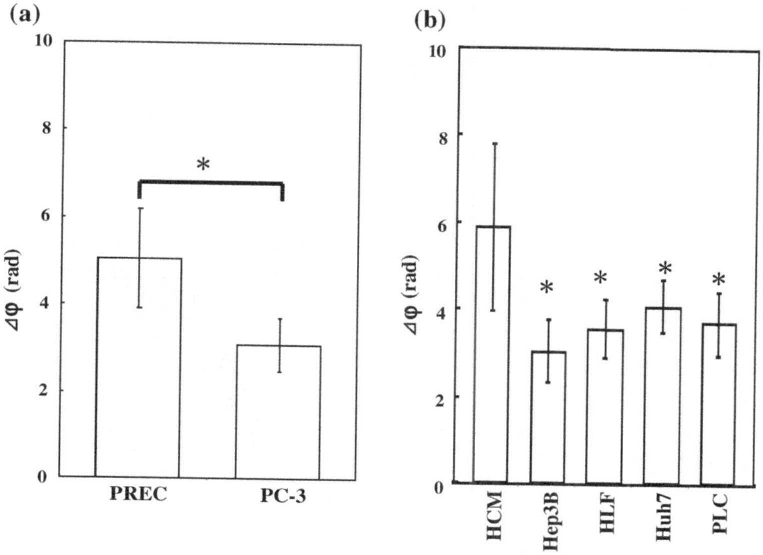

Fig. 8 Comparison of laser phase shifts of cancer and normal cells. Laser phase shift of PREC and PC-3 cells (**a**) and HCM and four kinds of human hepatocellular carcinoma cells (**b**) were determined by PLM. (average ± SD, *; $p < 0.01$) [22]

3.4.2 Correlation Between Actin Content and Phase Shift

However, it is not clear whether cytoskeleton content affects the difference in phase shift between cancer and normal cells and whether cell refractive index or cell height affects that difference. Consequently, the reason for the difference in phase shift between cancerous PC-3 cells and normal PRECs was investigated [26].

The cells attached to the bottom surface of the culture dish were fixed with 4 % paraformaldehyde, washed with PBS, and incubated with 0.1 % Triton X-100 in PBS for 5 min. They were washed with PBS and incubated with 25 times diluted rhodamine–phalloidin solution (0.56 μM, Cytoskeleton Inc., Denver, USA) for 45 min in the dark. They were washed three times with PBS and observed using a fluorescence microscope (IX70, Olympus, Tokyo, Japan) under a constant exposure condition (535-nm excitation filter and 585-nm adsorption filter). The line surrounding the area of a cell was drawn in the blight field, and the average fluorescent intensity for that area was calculated using Image J. Thus, average of fluorescence intensity in cell adhesion area was calculated as the actin content.

The actin content and phase shifts of adherent PREC and PC-3 cells with or without the cytochalasine D treatment (50 cells) were analyzed. As shown in Figs. 9a, b, the average fluorescence intensities as an index of actin densities of PREC and PC-3 cells with the treatment were lower than those of cells without the treatment. Also, the average phase shifts of PREC and PC-3 cells with the treatment were lower than those without the treatment. The plot of phase shift of all these cells

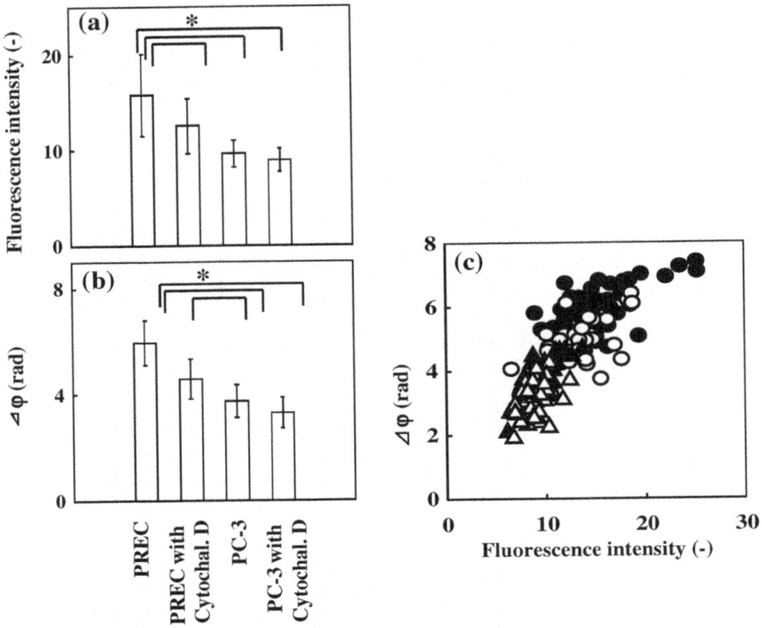

Fig. 9 Correlation between the phase shift and actin content of PRECs and PC-3 cells with or without cytochalasine D treatment

against fluorescence intensity showed a marked positive correlation (r = 0.84) between phase shift and actin content (Fig. 9c). Those findings showed that the phase shift decreased as actin content decreased in PRECs and PC-3 cells.

Atomic force microscopy (AFM) revealed that the elasticities of two kinds of normal human bladder epithelial cells (Hu609 and HCV29) are ten times higher than those of three kinds of cancerous cells (Hu456, T24, and BC3726) [20, 21]. Such elasticities were reported to partly correlate with the lower content of cytoskeleton in cancerous cells [27–29]. These reports support the lower actin content in cancerous PC-3 cells than in normal PRECs.

Both of the phase shift and actin content of cancerous PC-3 cells were markedly lower than those of normal PRECs, and there was a markedly positive correlation between the phase shift and actin content of these cells. The treatment of PREC and PC-3 cells with cytochalasin D decreased both of the phase shift and actin content, and there was a marked positive correlation between phase shift and actin content in all cell populations including cancerous PC-3 cells and normal PRECs with and without the treatment with cytochalasine D.

These findings and reports strongly suggest that the lower phase shift of cancerous PC-3 cells than of normal PRECs could be attributed to the lower actin content of PC-3 cells.

3.4.3 Time-Lapse Analysis of Phase Shift for Noninvasive Discrimination of Human Normal Cells and Malignant Tumor Cells

It was supposed that cancerous PC-3 cells in the interphase could be clearly discriminated from normal PRECs by the measurement of phase shift, if the laser phase shifts of PC-3 cells in the interphase are markedly lower than those of PREC cells in the interphase. In that case, it is possible that adherent PC-3 cancer cells could be noninvasively discriminated from normal adherent PRECs by the periodical measurement of phase shift using PLM. Consequently, the time-lapse analysis of phase shift during the whole cell cycle of PRECs and PC-3 cells was performed [30].

The PC-3 cells and PRECs were plated on a dish with grids (11.8 cm^2; SARSTEDT, Nümbrecht, Germany) at a density of 0.15×10^4 cells/cm^2 using their respective media and incubated in a conventional incubator at 37 °C in 5 % CO_2 atmosphere for 72 h, respectively. One dish containing cells was set in the small incubator on the observation stage of a PLM whose temperature and atmosphere were respectively controlled at 37 °C and 5 % CO_2. Two neighboring fields of view, with and without cells, were selected as the sample and reference fields, respectively. The recording of interference images between the sample and reference fields and the determination of the laser phase shift ($\Delta\varphi$) for all the pixels in the sample field were performed periodically by PLM at a constant interval (15–20 min). After every determination of phase shift, the sample field was observed under an inverted microscope and the copy of the bright field image shown on a display was taken using the "copy screen" function of a computer.

To confirm the change in the phase shift of cancerous PC-3 cells during culture, time-lapse analysis of PC-3 cells was performed. All images of the bright field, the interference, and 3D $\Delta\varphi$ showed the division of a cell to daughter cells from 10.0 to 10.3 h. These times of the mitotic phase are indicated by asterisks in Fig. 10b,

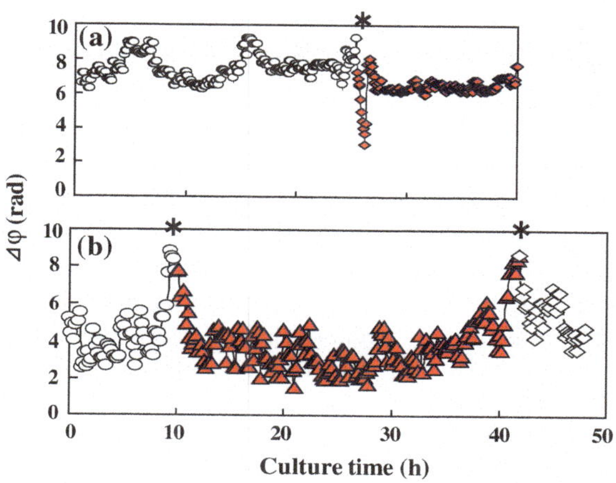

Fig. 10 Time-lapse observation of PC-3 and PREC cells

which shows the highest phase shift in each cell. The phase shift markedly increased more than 1.2 rad temporarily during the mitotic phase, while the phase shift was in the range between 1.5 and 6.2 rad during the interphase.

To confirm the change in phase shift of normal PRECs during culture, time-lapse analysis of PRECs was performed. All images of the bright field, the interference, and 3D $\Delta\varphi$ showed the division of cell daughter cells. The times of the mitotic phase are indicated by asterisks in Fig. 10a, which shows the highest phase shift in each cell. The phase shift markedly increased more than 1.2 rad temporarily during the mitotic phase, while the phase shift was in the range between 6.1 and 8.0 rad during the interphase.

Comparison of the time course of the phase shift of cancerous PC-3 cells (Fig. 10b) with that of normal PRECs (Fig. 10a) revealed that the phase shift of the PC-3 cell during the interphase was lower than the phase shift of the PREC except for the temporal decrease after the mitosis of the PREC. On the basis of these findings, a method to detect and destroy the contaminating cancer cells during cultivation was proposed. Briefly, the phase shift of all cells in a culture is periodically determined by PLM at intervals of 10 h, for example (approximately one-half of mean generation time). If the phase shift of certain cells is lower than a threshold (e.g., 6.0 rad), these cells can be destroyed by the irradiation using a femtosecond laser [31], although the effect of the irradiation on growth or death of the desirable normal cell is not yet clear. Even if the detection time was the mitotic phase of a PC-3 cell and the phase difference of the cancerous PC-3 cells was high (e.g., 7.5–9.0 rad), the phase shift of that PC-3 cell will become lower and that cell could be discriminated as a cancerous cell in the next detection time (e.g., 10 h later). Thus, almost all cancerous PC-3 cells will be discriminated as cancerous cells and destroyed within a few times of detection. In conclusion, it will be proposed that adherent cancer PC-3 cells could be noninvasively discriminated from normal adherent PRECs by the periodical measurement of phase shift using PLM, although a better method will be proposed in near future.

3.5 Chemical Analysis of Culture Components

3.5.1 Chemical Analysis of Culture Supernatant

In general, medium change, in which the culture supernatant is replaced with fresh medium, is performed during the culture of mammalian cells to remove accumulated toxic substances such as ammonia and lactate and supply growth factors and energy sources such as glucose and glutamine at appropriate concentrations. While the frequency of medium change may depend on the cell density and type of medium, medium change may be performed daily to weekly. Determination of the chemical composition of such culture supernatant may provide several types of useful information on the quality of cells under cultivation such as metabolic activity and differentiation level. The chemical analysis of the composition of the spent medium by conventional analytical methods such as ELISA and HPLC may be noninvasive for cells under cultivation.

Recently, a novel protein named MIA (melanoma inhibitory activity), derived from metastatic melanoma cell lines, has been characterized [32, 33]. In nonmalignant tissues, MIA expression is predominantly seen in developing and mature cartilage, whereas in malignancies, elevated serum levels are predominantly found in metastatic melanoma, and to a lesser extent in ovarian, pancreatic and breast cancer [34, 35]. MIA was suggested to be an important regulator of chondrogenic differentiation and maintenance of cartilage [36]. It was shown that measuring MIA in tissue culture supernatant by quantitative ELISA can be used as a marker for differentiated chondrocytes [37].

Indeed, it was reported that there was a marked positive correlation between aggrecan gene expression level and cell-specific production rate of MIA during the differentiation of human bone marrow MSCs to chondrocytes and redifferentiation of passaged human chondrocytes [38]. These results suggested that the level of differentiation of MSCs to chondrocytes could be noninvasively estimated by determining the MIA concentration in the culture supernatant, while only the average of the differentiation level of a cell population is available.

3.5.2 Chemical Analysis Using a Probe Beam

A simple noninvasive method of identifying a dead or live cell using a probe beam was reported [39]. If a cell is alive, the active transport of materials across the cell membrane causes a change in concentration gradients, and this change further induces a change in the deflection of a probe beam passing through the vicinity of the cell membrane. If a cell is dead, no or little change in the deflection of the probe beam is induced because there is no or little active movement of materials across the cell membrane.

As shown in Fig. 11a, a probe beam from a diode laser (output power 3 mW; wavelength 667 nm) was introduced into a microscope with a micro-XY-stage for adjusting the distance between the probe beam and a cell adhered on the bottom of a culture dish. The probe beam was reflected inside the microscope and then focused (spot size approx. 5 μm) onto the vicinity of the cell membrane of a cultured human cell using an objective lens (N.A. 0.1). The deflection of the probe beam is detected using a bicell photodiode. The deflection of the probe beam is monitored, and it was found that whether a cell is dead or alive as determined from the deflection signal agreed with the conventional decision.

3.5.3 Chemical Analysis Using Infrared Beam

An in situ monitoring system of live-cultured cells by infrared adsorption spectroscopy in the geometry of multiple internal reflections (MIR-IRAS) was reported [40, 41]. The volume of the sample solution in the equipment for MIR-IRAS

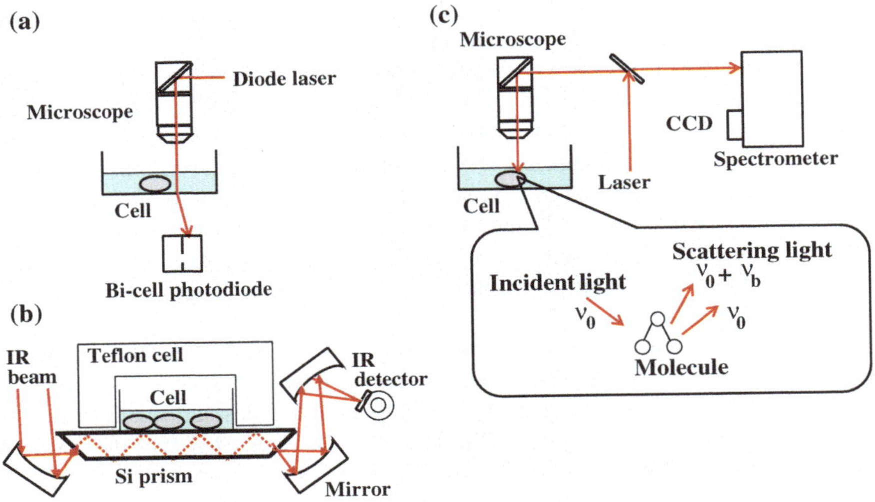

Fig. 11 Noninvasive chemical analysis of components of adherent mammalian cell culture. **a** Single-cell analysis using a probe beam, **b** MIR-IRAS measurement, **c** Raman spectroscopy

measurement was 100–200 μL (Fig. 11b). A custom-made Si prism was $0.5 \times 10 \times 30$ mm^3 with 45° bevels on each of the short edges and is in contact with the sample solution. An infrared light beam from an interferon-ester (BOMEM MB-100) was focused at normal incidence onto one of the two levels of the Si prism and penetrated through the Si prism, internally reflecting about 60 times. The light that propagated through the Si prism to the other bevel was focused onto a liquid-nitrogen-cooled Mercury–Cadmium–Telluride detector. The internal reflections that excited an evanescent field influenced the propagated light.

To observe live-cultured cells, the temperature of the sample chamber of a Fourier transform infrared (FTIR) spectrometer was maintained at 37 °C and a humidified gas mixture containing 5 % CO_2 was introduced into the sample chamber. MCF-7 cells, a human breast cancer cell line, cultured on Si MIR prisms were placed in the sample chamber, and the infrared spectra of MCF-7 cells were collected for 5 h. It was found that the adhesion and metabolism of MCF-7 cells could be monitored using the absorption intensity of the amide-II protein band (1,545 cm^{-1}) and also using the absorption intensities of CH_x bands (2,700–3,100 cm^{-1}). These results suggested that this system is useful for the noninvasive monitoring of cell viability, while the spatial resolution was not high and the cells needed to adhere onto the prism.

3.5.4 Raman Spectroscopy

Vibrational spectroscopy provides molecular-level information on the investigated sample and can be widely used for qualitative and quantitative analytical studies.

Raman spectroscopy is one such technique, and it measures the inelastic low-energy light scattering that is emitted when incident monochromatic light interacts with the sample (Fig. 11c). The change in energy correlates with the excitation of a vibrational mode in the molecule or part of a molecule. Thus, a Raman spectrum represents a chemical fingerprint of the molecular composition of the sample.

Raman spectroscopy has several advantages over its counterpart FTIR spectroscopy. Firstly, water has only a weak Raman spectrum; therefore, many functional groups can be observed in aqueous media and spectral information can be recorded from living systems. Secondly, in the case of biological samples, Raman spectra often exhibit a number of sharp bands, whereas the infrared spectra of cells and tissue often show broad spectral features. Thirdly, Raman microscopy offers a higher spatial resolution that cannot be reached even by using synchrotron IR sources.

Bio-Raman spectroscopy, combined with multivariate statistical analysis techniques, can be used to noninvasively identify biochemical differences in fetal osteoblasts (FOBs) exposed or not exposed to 45S5 Bioglass (BG)-conditioned media. Gene and protein expression studies were also performed for known markers of osteoblastic differentiation, namely alkaline phosphatase (ALP),bone sialoprotein (BSP), and collagen type I (Col I). Quantitative RT-PCR confirmed the upregulation of genes associated with osteoblast differentiation after exposure to BG-conditioned media. These results suggested that Raman spectroscopy can be used to noninvasively detect biochemical changes in FOBs associated with differentiation [42].

Calu-1 lung cancer cells were used as a model, and their morphology and growth were assessed on the Raman substrate of quartz. Forty cells were analyzed, and Raman spectra were collected from nuclei and cytoplasms using a 785-nm laser excitation of 30 mW, in the spectral range of 580–1750 cm^{-1}, and an acquisition time of 2 × 10 s/spectrum. Discriminant spectral information related to the nucleus and cytoplasm was extracted by multivariate statistical methods and attributed to nucleic acids, lipids, and proteins. Finally, Raman spectral imaging was performed to show the distribution of these components within the cell [43].

Chondrocytes isolated from calf articular cartilage from multiple passages (P0–P4) were seeded onto coated CaF_2 slides and incubated overnight at 37 °C under 5 % CO_2 atmosphere. The chondrocytes showed decreases in the gene expression levels of chondrogenic differentiation markers (collagen type II, aggrecan, SOX 9) after multiple passages, and Raman spectra (647.1 nm) of single chondrocytes following multiple passages showed a decrease in the protein and nucleic acid bands [44]. Chondrocytes could be noninvasively discriminated from fibroblast cells by Raman spectroscopy [45].

4 Conclusion and Perspective

The condition of development of cell processing engineering was reviewed, while there may be several subjects and examples which were not introduced here. Please refer some other reviews on the same subject.

Along with the increasing numbers of tests under clinical trial and the products of regenerative medicine, the demand of automatic cell processing should be growing. While the rule and guideline for the quality of automatic cell processing machine may not be enough, several kinds of machine which are respectively applicable to several types of cell processing including three-dimensional cultures will appear in future. The development of allogeneic transplantation system may also affect the needs of automatic cell processing machine.

There should be several kinds of analytical methods which are not yet applied to noninvasive estimation of cell quality. One of the reasons may be no development of such an analytical machine version for living cell analysis. There is also little machine for the observation of three-dimensional culture and tissue.

Noninvasive estimation of cell quality may not be essential for allogeneic system. However, noninvasive method should be important also for allogeneic system, because almost noninvasive methods are automatic. Moreover, noninvasive estimation of cell quality is also applicable to cultivation process control during the culture. The noninvasive estimation of cell quality may have needs in other clinical field than regenerative medicine like clinical diagnosis.

References

1. Yoshida T, Takagi M (2004) Cell processing engineering for ex vivo expansion of hematopoietic cells. Biochem Eng J 20(2–3):99–106
2. Takagi M (2005) Cell processing engineering for ex-vivo expansion of hematopoietic cells: a review. J Biosci Bioeng 99(3):189–196
3. Nakajima K, Kanazawa K, Takagi M, Wakitani S, Inagi M (2009) Development of the automatic cell culture machine for the adherent cell. Inflam Regen 29(2):133–136
4. Yokoyama M, Miwa H, Maeda S, Wakitani S, Takagi M (2008) Influence of fetal calf serum on differentiation of mesenchymal stem cells during expansion. J Biosci Bioeng 106(1):46–50
5. Takagi M, Kitabayashi T, Koizumi S, Hirose H, Kondo S, Fujiwara M, Ueno K, Misawa H, Hosokawa Y, Masuhara H, Wakitani S (2008) Correlation between cell morphology and aggrecan gene expression level during differentiation from mesenchymal stem cells to chondrocytes. Biotechnol Lett 30:1189–1195
6. Lee CR, Grodzinsky AJ, Spector M (2003) Modulation of the contractile and biosynthetic activity of chondrocytes seeded in collagen-glycosaminoglycan matrices. Tissue Eng 9 (1):27–36
7. Oda R, Suardita K, Fujimoto K, Pan H, Yan W, Shimazu A, Shintani H, Kato Y (2003) Anti-membrane-bound transferrin-like protein antibodies induce cell-shape change and chondrocyte differentiation in the presence or absence of concanavalin. Am J Cell Sci 116:2029–2038
8. Li K, Miller ED, Chen M, Kanade T, Weiss LE, Campbell PG (2008) Cell population tracking and lineage construction with spatiotemporal context. Med Imag Anal 12:546–566
9. Kato R, Shiono H, Yamamoto W, Nagura Y, Mukaiyama K, Kojima K, Kii H, Koshiba R, Yamada A, Uozumi T, Watanabe H, Mizuno J, Tomioka K, Honda H (2009) Cell quality prediction system based on image analysis combined with bioinformatics for the quality control of regenerative cell therapy. Tissue Eng Regen Med 6(12):S95
10. Kagalwala F, Kanade T (2003) Reconstructing specimens using DIC microscope images. Cybernetics 33(5):728–737

11. Fotiadis D, Scheuring S, Muller SA, Engel A, Muller DJ (2002) Review: imaging and manipulation of biological structures with AFM. Micron 33:385–397
12. Takagi M, Hayashi H, Yoshida T (2000) The effect of osmolarity on metabolism and morphology in adhesion and suspension Chinese hamster ovary cells producing tissue plasminogen activator. Cytotechnology 32:171–179
13. Kemper B, Carl D, Schnekenburger J, Bredebusch I, Schäfer M, Domschke W, von Bally G (2006) Investigation of living pancreas tumor cells by digital holographic microscopy. J Biomed Opt 11:34005
14. Endo J, Chen J, Kobayashi D, Wada Y, Fujita H (2002) Transmission laser microscope using the phase-shifting technique and its application to measurement of optical waveguides. Appl Opt 41(7):1308–1314
15. Takagi M, Kitabayashi T, Ito S, Fujiwara M, Tokuda A (2007) Noninvasive measurement of three-dimensional morphology of adhered Chinese hamster ovary cells employing phase-shifting laser microscope. J Biomed Opt 12(5):054010-1–054010-5
16. Sanger JW, Sanger JM (1980) Surface and shape changes during cell division. Cell Tissue Res 209:177–186
17. Ito S, Takagi M (2009) Correlation between cell cycle phase of adherent Chinese hamster ovary cells and laser phase shift determined by phase-shifting laser microscopy. Biotechnol Lett 31:39–42
18. Tokumitsu A, Wakitani S, Takagi M (2009) Noninvasive estimation of cell cycle phase and proliferation rate of human mesenchymal stem cells by phase-shifting laser microscopy. Cytotechnol 59(3):161–167
19. Hilliard C, Hill R, Armstrong M, Fleckenstein C, Crowley J, Freelamd E, Duffy D, Galloway S (2007) Chromosome aberrations in Chinese hamster and human cells: a comparison using compounds with various genotoxicity profiles. Mutat Res 616:103–118
20. Cross SE, Jin YS, Rao J, Gimzewski J (2007) Nanomechanical analysis of cells from cancer patients. Nat Nanotechnol 12:780–783
21. Lekka M, Laidler P, Gil D, Lekki J, Stachura Z, Hrynkiewicz AZ (1999) Elasticity of normal and cancerous human bladder cells studied by scanning force microscopy. Eur Biophys J 28:312–316
22. Tokumitsu A, Wakitani S, Takagi M (2010) Noninvasive discrimination of human normal cells and malignant tumor cells by phase-shifting laser microscopy. J Biosci Bioeng 109:499–503
23. Macnab GM, Alexander JJ, Lecatsas G, Bey EM, Urbanowicz JM (1976) Hepatitis B surface antigen produced by a human hepatoma cell line. Br J Cancer 34:509–515
24. Doi I, Namba M, Sato J (1975) Establishment and some biological characteristics of human hepatoma cell lines. Gann 66:385–392
25. Nakabayashi H, Taketa K, Miyano K, Yamane T, Sato J (1982) Growth of human hepatoma cell lines with differentiated functions in chemically defined medium. Cancer Res 42:3858–3863
26. Takagi M, Tokunaga N (2013) Correlation between actin content and phase-shift of adhesive normal and malignant prostate epithelial cells J Biosci Bioeng 115(3):310–313
27. Yamazaki D, Kurisu S, Takenawa T (2005) Regulation of cancer cell motility through actin reorganization. Cancer Sci 96:379–386
28. Li QS, Lee GYH, Ong NC, Lim TC (2008) AFM indentation study of breast cancer cells. Biochem Biophys Res Commun 374:609–613
29. Kim YC, Park SJ, Park JK (2008) Biomechanical analysis of cancerous and normal cells based on bulge generation in a microfluidic device. Analyst 133:1432–1439
30. Takagi M Shibaki Y (2012).Time-lapse analysis of laser phase shift for noninvasive discrimination of human normal cells and malignant tumor cells. J Biosci Bioeng 114 (5):556–559
31. Sakai J, Roldán D, Ueno K, Misawa H, Hosokawa Y, Iino T, Wakitani S, Takagi M (2012) Effect of the distance between adherent mesenchymal stem cell and the focus of irradiation of femtosecond laser on cell replication capacity. Cytotechnol. doi:10.1007/s10616-012-9437-2

32. Koehler MR, Bosserhoff AK, von Beust G, Bauer A, Blesch A, Buettner R, Schlegel J, Bogdahn U, Schmid M (1996) Assignment of the human melanoma inhibitory activity gene (MIA) to 19q13.32-q13.33 by fluorescence in situ hybridization (FISH). Genomics 35:265–267
33. Bosserhoff AK, Hein R, Bogdahn U, Buettner R (1996) Structre and promoter analysis of the gene encoding the human melanoma-inhibiting protein MIA. J Biol Chem 271:490–495
34. Bosserhoff AK, Kondo S, Moser M, Dietz UH, Copeland NG, Gilbert DJ, Jenkins NA, Buettner R, Sandell LJ (1997) Mouse CD-RAP/MIA gene: structure, chromosomal localization, and expression in cartilage and chondrosarcoma. Dev Dynam 208:516–525
35. Bosserhoff AK, Kaufmann M, Kaluza B, Bartke I, Zirngibl H, Hein R, Stolz W, Buettner R (1997) Melanoma inhibiting activity, a novel serum marker for progression of malignant melanoma. Cancer Res 57:3149–3153
36. Tscheudschilsuren G, Bosserhoff AK, Schlegel J, Vollmer D, Anton A, Alt V, Schnettler R, Brandt J, Proetzel G (2006) Regulation of mesenchymal stem cell and chondrocyte differentiation by MIA. Exp Cell Res 312:63–72
37. Bosserhoff AK, Buettner R (2003) Establishing the protein MIA (melanoma inhibitory activity) as a marker for chondrocyte differentiation. Biomaterials 24:3229–3234
38. Onoue K, Kusubashi H, Wakitani S, Takagi M (2009) Noninvasive estimation of aggrecan gene expression level of chondrocytes by the analysis of culture supernatant. Abstract of 61st annual meeting of the Japanese society for bioscience and bioengineering, Nagoya, Japan, 23–25 Sept
39. Wu XZ, Terada S (2005) Noninvasive diagnosis of a single cell with a probe beam. Biotechnol Prog 21:1772–1774
40. Miyamoto K, Yamada P, Yamaguchi R, Muto T, Hirano A, Kimura Y, Niwano M, Isoda H (2007) In situ observation of cell adhesion and metabolism using surface infrared spectroscopy. Cytotechnol 55:143–149
41. Yamaguchi R, Hirano-Iwata A, Kimura Y, Niwano M, Miyamoto K, Isoda H, Miyazaki H (2009) In situ real-time monitoring of apoptosis on leukemia cells by surface infrared spectroscopy. J Appl Phys 105:024701
42. Jell G, Notingher I, Tsigkou O, Notingher P, Polak JM, Hench LL, Stevens MM (2008) Bioactive glass-induced osteoblast differentiation: a noninvasive spectroscopic study. J Biomed Mater Res 86A:31–40
43. Draux F, Jeannesson P, Beljebbar A, Tfayli A, Fourre N, Manfait M, Sule-Suso J, Sockalingum GD (2009) Raman spectral imaging of single living cancer cells: a preliminary study. Analyst 134:542–548
44. Kunstar A, Otto C, van Blitterswij CA, van Apeldoorn AA (2009) Raman monitoring of chondrocyte dedifferentiation and differentiation. Tissue Eng Regen Med 6(12):S231
45. Takagi M, Miyata Y, Wakitani S, Ishizaka, S, Kitamura N (2009) Noninvasive discrimination of adherent chondrocytes from fibroblasts using Raman spectroscopy. Abstract of 61st annual meeting of the Japanese society for bioscience and bioengineering, Nagoya, Japan, 23–25 Sept

Adv Biochem Eng Biotechnol (2016) 152: 75–90
DOI 10.1007/10_2015_307

Published Online: 11 March 2015

Reactors for High Solid Loading Pretreatment of Lignocellulosic Biomass

Jian Zhang, Weiliang Hou and Jie Bao

Abstract The review summarized the types, the geometry, and the design principle of pretreatment reactors at high solid loading of lignocellulose material. Among the reactors used, the explosion reactors and the helical stirring reactors are to be considered as the practical form for high solids loading pretreatment operation; the comminution reactors and the extruder reactors are difficult to be used as an independent unit, but possible to be used in the combined form with other types of reactors. The principles of the pretreatment reactor design at high solid loading were discussed and several basic principles for the design were proposed. This review provided useful information for choosing the reactor types and designing the geometry of pretreatment operation at the high solids loading.

Keyword Lignocellulosic biomass · High solid loading · Pretreatment reactors · Mixing · Cellulose conversion

Contents

1 Background 76
2 Reactors for Different Pretreatment Operations at High Solid Loading 77
2.1 Comminution Equipment 77
2.2 Extruder Reactors 78
2.3 Explosion Reactors 81
2.4 Helical Ribbon Stirring Pretreatment Reactors 83
3 Design Principles of Pretreatment Reactors Operated at High Solid Loadings 84
References 86

J. Zhang · W. Hou · J. Bao (✉)
State Key Laboratory of Bioreactor Engineering,
East China University of Science and Technology,
130 Meilong Road, 200237 Shanghai, China
e-mail: jbao@ecust.edu.cn

1 Background

Pretreatment is a crucial step for overcoming the recalcitrance of lignocellulosic biomass in biorefinery processing [1–7]. It also affects both upstream and downstream operations beyond the pretreatment itself [8]. The efficiency of a biorefinery process is highly dependent on pretreatment technology used, such as pretreatment method, severity, inhibitors generated, wastewater discharge, residues released, and the reactors used. The current leading pretreatment technologies include dilute acid, steam explosion, ammonia fiber expansion (AFEX), liquid hot water, flow-through acid, ammonia recycle percolation (ARP), lime pretreatment, and cellulose solvent-based pretreatments etc. [9–11]. For easy operation, low solid loading with the ratio of the solid biomass to the liquid solution below 1:5 is usually preferred by most pretreatments. However, low solid loading raises several serious problems in practice: (1) Massive amount of wastewater is generated as a heavy environmental burden because the wastewater contains considerable amount of monomer and oligomer sugars, as well as various chemical substances degraded from lignocellulose biomass; (2) a considerable amount of fine feedstock solid particles and fermentable sugars are lost in the consequent solid–liquid separation, which decreases overall yields of sugars; (3) cellulose, hemicellulose, and lignin are easy to be degraded into inhibitory compounds such as hydroxymethylfurfural (HMF), furfural, acetic acid, phenolic aldehydes, or acids in the aqueous environment; and (4) high capital investment due to low processing capacity of pretreatment reactors. High solid loading in biomass pretreatment becomes a reasonable option and trend in the future industrial pretreatment operation [12–14].

To obtain an ethanol-rich broth (beer broth) with a titer of more than 5 % (w/w) as the feed stream of the first distillation column, the solid loading in the enzymatic hydrolysis and fermentation step should be greater than 20 % (w/w) at a moderate ethanol yield of 0.45 g/g cellulose [15–17]. Thus, the solid loading in pretreatment should be at least greater than this value. In this review, the high solid loading pretreatment is defined as the operation with the solid loading higher than 20 % (w/w). At such a high solid loading, the mixing and transfer phenomena are critically important for maintaining the satisfactory pretreatment efficiency. These phenomena include the following: (1) the mass transfer of small amount of liquid (either hot water, alkaline solution, or dilute acid solution) among and within the large bulk of lignocellulosic biomass particles, (2) the heat transfer from small amount of hot steam to large bulk of solid lignocellulosic biomass, and (3) the massive volume shrinkage of the solid lignocellulose coming across the hot steam (especially as for the herbaceous biomass). To achieve a satisfactory mixing and transfer condition, appropriate design of pretreatment reactors is required. In this review, recent advances in the high solid loading pretreatment are reviewed with the focus on the reactor design for achieving the practical pretreatment operation.

In spite of the potential benefits from high solid loading pretreatment operations, difficulties in reactor design and operation limit intensive investigations. To our knowledge, only limited numbers of pretreatment reactors have been designed for the high solid loading operations. In this review, the pretreatment reactors suitable for high solid loading operations are summarized and analyzed. Four major reactor types are included according to their different configurations: comminution equipment, extruder reactors, explosion reactors, and helical stirring reactors.

2 Reactors for Different Pretreatment Operations at High Solid Loading

2.1 Comminution Equipment

Comminution is a common mechanical operation such as chopping, shredding, grinding, and milling by reducing the particle size and increasing the available specific surface areas, in which the degree of polymerization (DP) and cellulose crystallinity of lignocellulosic biomass could be reduced [1, 6, 18, 19]. Such a size reduction method is similar to the woody fiber production [20]. It does not involve any chemical reactions, and there is no fermentable components loss and inhibitor generation [21]. Comminution is generally operated continuously at a dry mode (solid loading up to 90 %, w/w). Hammer mills, disk mills, and ball mills are the major comminution equipment as shown in Fig. 1 [21, 22].

Hammer milling and disk milling are the two predominant methods to reduce the particle size of lignocellulose. Hammer milling reduces particle size of solid biomass by shear and impact actions. The hammer mill machine consists of several swinging hammers attached to a shaft and powered by an electric motor. The perforated metal screens covering the discharge opening of the mill retain coarse materials for further grinding while allowing the properly sized materials to pass as finished product [23, 24].

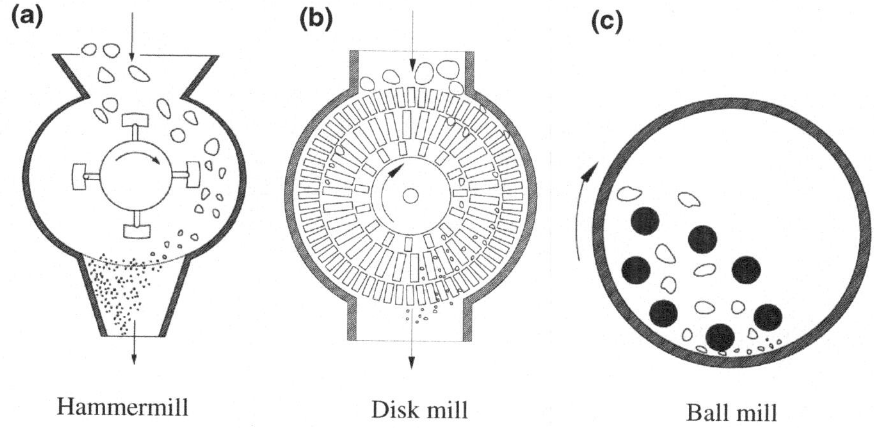

Fig. 1 Schematic diagram of the three representative comminution equipment. **a** Hammer mill. **b** Disk mill. **c** Ball mill

Disk mills are media mills that are built in a single- or double-disk version with both straight and profiled blades. Material is fed through a central orifice coaxially with the rotation axis and then is milled by a number of rotating blades. Material is fragmented between disks and flows to their periphery. Other disk mills perform the final fragmentation of material between the rotating disk and a ring-like stator coaxially with the rotor. Shear and compression are the predominant comminution mechanisms in these types of mills [25]. In addition, disk milling has been mainly used for wood fiber production and considered as one of the best fiberization processes for woody feedstock in commercial fiber production. While as for the lignocellulose biorefinery technology, disk milling is usually used as a post-pre-treatment operation coupled with other chemical pretreatment methods enhancing the pretreatment efficiency [26, 27].

Ball mill is configured by a cylinder containing numbers of balls and mounted on a disk driven by an alternating current dynamo. Revolution begins when the disk moves; meanwhile, the cylinders rotate around the axis of the disk in the reverse direction. The balls and solids collide with each other with centrifugal force, making the solids grind to small pieces. The ball milling can be combined with other pretreatment operations to improve the cellulose conversion yield [21, 28, 29].

The above three types of size reduction equipment are used either singly or in combination with other pretreatment steps [21, 26, 28–31]. The major problem of the comminution equipment is the intensive energy requirement [10, 32, 33]. A recent study showed that the particle size change of the raw corn stover did not alter the enzymatic hydrolysis yield in a certain range such as 0.2–2 mm [34]. According to Silva et al. [21], only the particle size less than 100 μm, specifically at 20 μm, had a dramatic facilitating effect on the cellulose hydrolysis, while this size required great energy input. The energy consumption for herbaceous biomass comminution to 0.3–3-mm particles is approximately 45 kWh/ton (dry) [10, 23, 24]. For the woody biomass, the energy could be one order of magnitude greater up to 500–800 kWh/ton (dry) [22, 35]. Therefore, comminution pretreatment might work only for herbaceous lignocellulosic feedstock, rather than the woody biomass.

It was also found that post-milling of chemically pretreated lignocellulosic biomass could enhance the enzymatic hydrolysis by over 90 %; meanwhile, the energy consumption on comminution was reduced by 80 % [30–32, 36, 37]. Therefore, a combination of comminution with other pretreatment methods, such as chemical or physical–chemical pretreatment, aqueous swelling, and chemical augmentation of milling etc, could be practically more attractive than sole milling of raw lignocellulose feedstock [29, 38–40].

2.2 Extruder Reactors

Extruder is widely used in plastics and food industries [41]. Due to its features of high shear force and easy operation control, extrusion has been introduced into the lignocellulose biorefinery as a pretreatment method [42, 43]. The detailed mechanism of extrusion pretreatment has not been well understood yet. Studies suggested that frictional and shearing forces might cause the fibrillation and shortening of biomass fibers. The heat generated by extrusion in the screw barrel can elevate the temperature to about 99 °C without additional heat and can facilitate the solid–liquid mixing and reaction [44]. These two effects partially damage the cellulose structure of lignocellulose and increase the specific surface area, thus enhancing the enzymatic hydrolysis performance.

The extruders used for pretreatment has two different types, single-screw extruder and twin-screw extruder [45, 46]. A typical extruder is usually composed of barrels, shafts and screws, material feeding system, and drive motor. The screw is the key part of the extruder. The single-screw and the twin-screw extruders are shown in Fig. 2. Forward conveying screws (FCS), forward kneading screws

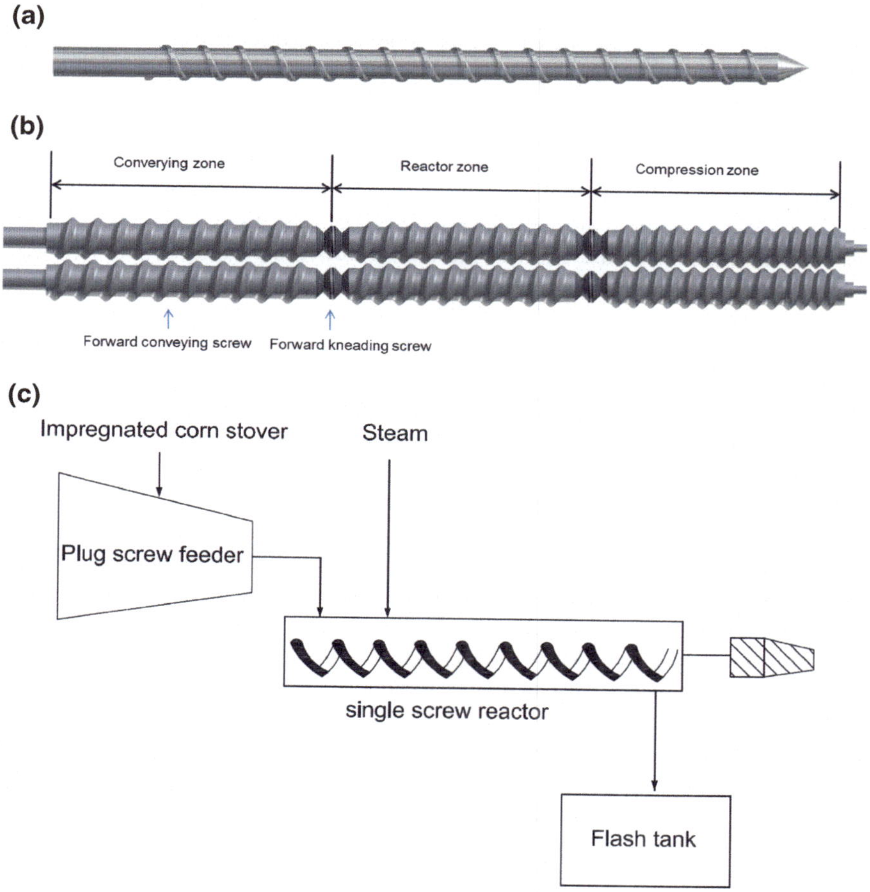

Fig. 2 Schematic diagram of the extruder screw configuration. **a** Single-screw extruder (adapted from Karunanithy et al. [48]). **b** Twin-screw extruder (adapted from Yoo et al. [49]). **c** Pilot-scale dilute H_2SO_4 pretreatment system in NREL (adapted from Humbird et al. [54])

(FKS), and the shortest pitch screws are the three main screw types. When the extruders are used for pretreatment, a catalyst (liquid acid, base, or surfactant) feeding apparatus is also included [45, 47, 48]. Lignocellulosic biomass is fed into the pretreatment section of the extruder by a hopper. The first several screws enable continuous movement of material toward the kneading block, where liquid catalyst is added if needed. FKS are installed to achieve good pulverization. Then, FCS and FKS are installed repeatedly with shorter pitch lengths for well mixing and longer reaction time. The last several screws, with the shortest pitch length, are used to create more pressure on the pretreatment [49, 50]. Screw unloading may generate limited explosion effect [51].

The extrusion pretreatment at high solid loading has been proved to enhance the enzymatic hydrolysis of lignocellulosic biomass. A combination of the twin-screw extrusion and the acid-catalyzed hot water extraction process was tested by Chen et al. [50, 51]. The extrusion treatment was conducted at approximately 50 % (w/w) solid loading of rice straw, and the xylan conversion ratio reached 83.7 % after pretreated at the optimum pretreatment condition. The cellulose-to-glucose yield of 80 % after enzymatic hydrolysis was obtained [50, 51]. However, an extra energy-intensive hot water extraction and solid/liquid separation step were needed for this biorefinery scenario. Zhang et al. obtained a sugar recovery of as high as 91.07 % with the twin-screw extrusion pretreatment of corn stover at a biomass/liquid ratio of 1:2, while the cellulase dosage was a little higher (24 FPU/g substrate) [52, 53]. Karunanithy et al. tested the effect of moisture content of pine wood chips (25, 35, and 45 % wet base moisture content) on the extrusion efficiency and reached an enzymatic hydrolysis yield of 66.1 % [47, 48]. Zhang et al. [43] found that the optimum glucose, xylose, and combined sugar recoveries from corn stover were 48.79, 24.98, and 40.07 %, respectively, at 27.5 % moisture content using the twin-screw extrusion pretreatment. Yoo et al. [49] found that the greater glucose yields (higher than 70 %) were observed at higher in-barrel moisture (45 and 50 %) and lower screw speed (280 and 350 rpm) in the thermo-mechanical extrusion pre-treatment of soybean hulls. Although the above studies worked well with the extruder equipment, glucose yield showed a little lower compared with the other chemical or physical–chemical pretreatments. In addition, extrusion pretreatment was effective for various feedstocks including corn stover, pine wood chips, and soybean hulls. Chemical addition of alkaline or acid could enhance extrusion pretreatment efficiency [40, 44, 45, 53].

A pretreatment design by National Renewable Energy Laboratory in 2011 (NREL Technical report TP-5100-47764, 2011) includes a feedstock receiving system, a vertical vessel for steam-heating and a horizontal extrusion pretreatment reactor (Fig. 2c) [54]. The extrusion reactor is operated at a higher pressure and a short residence time. High-pressure steam is injected into this vessel to maintain temperature. The horizontal reactor configuration is chosen here because it allows tighter residence time distribution control than a vertical reactor. The final xylan and glucan conversions after enzymatic hydrolysis were higher than 85 and 90 %, respectively.

The extrusion pretreatment can be conducted continuously at a solid loading higher than 30 %; thus, wastewater generation and cellulose loss could be reduced, and the formation of inhibitors is also depressed. The extrusion pretreatment can be easily combined with other pretreatments, such as chemical or thermal–chemical methods, to further enhance the pretreatment efficiency. Furthermore, extruders are considered readily scalable equipment and have well-established vendor quotes. The disadvantages of extrusion pretreatment are summarized below:

1. Energy intensive: For example, the mechanical energy input for wheat bran extrusion treatment was 800–2,300 kJ/kg, and the sugar yield was only 41–60 %. When chemicals were added, the mechanical energy requirement could be reduced to 240–540 kJ/kg [55]. The energy consumption is close to or even higher than that for the comminution.
2. Lack of elasticity and flow ability of biomass suspensions at relatively high solid loading precludes its practical application. Incorporation of gelation agents such as starch, carboxymethyl cellulose, or surfactant into extrusion process is required [42, 56–58].
3. Low cellulose digestibility. Usually, acid or base additions are required to enhance the pretreatment efficiency.
4. Serious abrasion of extruders by lignocellulose materials increases the capital investment.

2.3 Explosion Reactors

Steam explosion was introduced and patented as a biomass pretreatment for wood pulp production in 1926 and then used for bioethanol or biogas production [59]. Explosion reactors and variants are commonly used for explosion-based pretreatment, such as steam explosion pretreatment, AFEX, and ARP [60–63]. The configuration of explosion reactors is simple (Fig. 3), including a pressure vessel, a steam (or other chemicals) inlet, and a fast-open release valve. Steam explosion pretreatment can be conducted at a solid loading of 20–50 %. After lignocellulosic biomass is loaded, the reactor is heated up to desired temperature and pressure, and then, the biomass materials are quickly discharged within 1/100 s to initiate an "explosion" by releasing the overheated liquid water into hot vapor. The rapid thermal expansion breaks lignocellulose structure and leads to crackdown of lignin shell and decrystallization of cellulose into amorphous cellulose. In addition, acetic acid released from the hydration of acetyl group catalyzes the partial degradation of hemicellulose during steam explosion pretreatment. Generally, steam explosion uses relatively large chip size to avoid the energy-intensive size reduction of biomass like grinding [64].

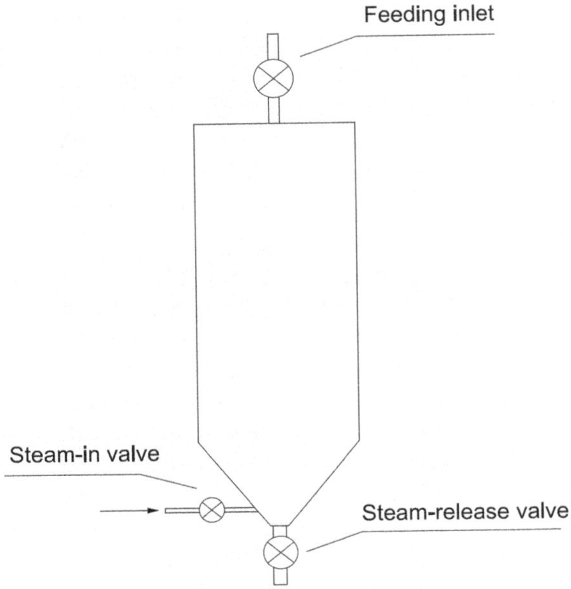

Fig. 3 Schematic diagram of explosion pretreatment reactor

Addition of acids (sulfuric acid, SO_2, or CO_2) or alkaline reagents (ammonia or liquid ammonia) during the explosion pretreatment can greatly enhance the pretreatment efficiency or shorten the process time and temperature. Linde et al. [65] found that the steam pretreatment of wheat straw gained satisfactory results (glucose and xylose recovery of 102 and 96 %, respectively) when impregnated with 0.2 % sulfuric acid for 1 h and then pretreated at 30 % dry solid loading at 190 °C for 10 min before explosion. When sugarcane bagasse was used, the overall sugar yield reached 87 % after impregnation with 2 % SO_2 for 30 min and then pretreatment at 25 % solid loading at 190 °C for 5 min before explosion [66]. When supercritical carbon dioxide was added to the steam explosion pretreatment of corn stover, the glucose yield was only 30 % at the extremely high pressure of 25 MPa and 150 °C at 25 % solid loading for 1 h before explosion [63]. AFEX was usually carried out with the ammonia loading of 1.0 g/g dry biomass and water loading of 0.6 g/g dry biomass, at 90–140 °C for 10–15 min for obtaining 90 % of the total sugar yield [67–74]. Kim and Lee developed an ARP pretreatment process and achieved 88.5 % digestibility of the corn stover when pretreated at 170 °C with 3.3 mL of 15 % NH_3 per gram of corn stover at a flow rate of 5.5 mL/min [62].

Mixing is not suitable for explosion reactors because of obvious damage on the rotating impellers by explosion actions. Therefore, sufficient impregnation of the lignocellulose feedstock with liquid ammonia or other reagents is usually performed at a low solid loading, and then, a solid/liquid separation unit is used to squeeze the extra liquid out before employing the high solid loading pretreatment. The severe explosion pretreatment conditions, including considerably high temperature and pressure, explosion effect, and the reagents added, propose a tough requirement for reactor materials, sealing parts, installation, and operation. The maintenance and lifetime of explosion reactor are also to be the major concern. The recycling of the catalysts (NH_3) is an expensive operation. Besides, machining accuracy and fine design of the explosion reactor for continuous operation is still a great challenge for researchers.

2.4 Helical Ribbon Stirring Pretreatment Reactors

Helical ribbon reactor is used for mixing the high viscous fluids in petrochemical industry and pulp manufacture in paper industry, but the application to lignocellulose biorefinery process at high solid loading is a novel idea. At first, the helical stirring reactor was developed to solve the mixing problem during simultaneous saccharification and ethanol fermentation (SSF) at high solid loading (30 %, w/w) for obtaining a high final ethanol titer [75]. Later on, the helical ribbon reactor was applied to the dilute acid pretreatment at very high solid loading up to 70 % (w/w). The configuration of a typical helical ribbon pretreatment reactor is shown in Fig. 4. It includes a helical ribbon stirrer mounted on the shaft and a bottom anchor stirrer to avoid sedimentation of the solid lignocellulose particles [14].

When a reactor for dry dilute acid pretreatment at nearly 70 % solid loading is scaled-up for industrial use, the cylinder reactor is unable to provide a good mixing of solids and the jetting steam unless a forced stirring mechanism is introduced. He et al. designed a helical ribbon stirrer similar to that in the bioreactor used for the dry dilute acid pretreatment reactor in light of its ability to deal with the high solid loading during SSF [14, 75]. The computational fluid dynamics (CFD) simulation showed that mixing between hot steam and the solid materials was greatly enhanced compared to that in an empty cylinder reactor [76]. The pretreatment efficiency was significantly improved, and the inhibitor generation was reduced dramatically. It was also worth noting that the labor-intensive presoaking operation before pretreatment to ensure the thorough impregnation of the lignocellulose by the liquid acid can be completely removed in the newly designed pretreatment reactor. The solid feedstock and the liquid acid can be fed into the reactor co-currently by helically agitation, and the pretreatment efficiency was same to that using the thoroughly presoaked feedstock [77].

Helical stirring pretreatment reactor fits the pretreatment operation well at high solid loading of lignocellulose. It can successfully avoid the labor-intensive presoaking step, the serious corrosion of the pretreatment reactor, and the massive pretreatment liquid generation. Currently, helical stirring reactor can only be operated at a batch mode, which may decrease the pretreatment productivity.

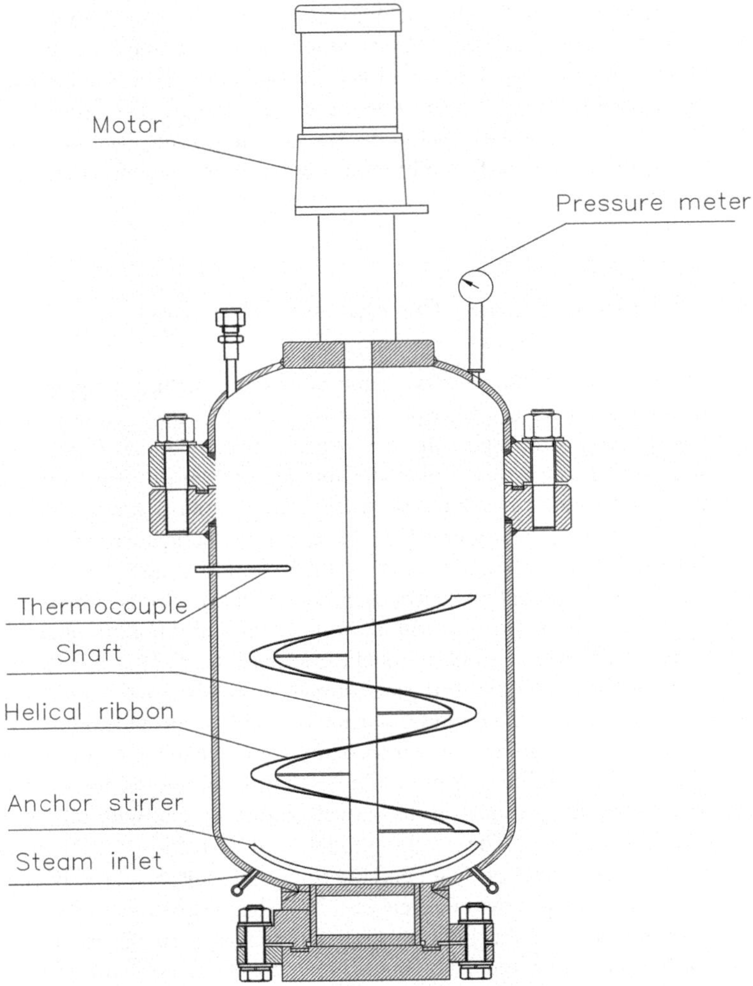

Fig. 4 Schematic diagram of pretreatment reactor with the helical ribbon stirring [14]

3 Design Principles of Pretreatment Reactors Operated at High Solid Loadings

When designing pretreatment reactors that can handle high solid loading of lignocellulose, it is technically challenging to achieve efficient mixing of solid with liquid and gaseous reagents as well as the transfer of heat and mass. Based on the features and performances of the four types of reactors used in different pretreatment methods discussed above, at least three fundamental requirements should be met for a successful design and operation of pretreatment reactors operate at high solid loading:

1. Low-energy input is prerequisite. The intensive energy consumption of comminution, screw extruder, and explosion reactors will definitely increase the final product costs and make this biorefinery scenario expensive and unpractical. Therefore, the energy consumption should be reduced to acceptable levels.
2. Satisfactory pretreatment efficiency at low severity. Low-severity pretreatment means low chemical loading and moderate temperature and pressure. This will impose less stress on the reactor, reduce the damage and corrosion to the reactor, reduce the sugar loss especially the hemicellulose derived sugars, and decrease the reactor manufacturing costs, maintenance as well as prolong their lifetime. Meanwhile, less inhibitor generation at a low pretreatment intensity will benefit for the downstream processing.
3. Continuous operation is desirable. Productivity is a key index for the industrial process especially for processing large amount of feedstock, which is often neglected in the research at a bench scale. Comminution, screw extruder, and explosion reactors can work continuously and be scaled-up easily, while the helical ribbon reactors are still challenging when applied to a large scale.

High solid loading leads to poor mixing between bulky solid biomass and small amount of liquid and vapor, creating significant difficulties in the pretreatment reactor design and operations. Different from the mature chemical reactor design methodology used in petrochemical industry, design and optimization of the pretreatment reactors is still a challenging job because the knowledge is still limited of rheological properties of lignocellulosic biomass, mechanisms of pretreatment operation, and mass and heat transfer from steam and hot water to solid lignocellulose [78–80].

When designing an efficient pretreatment reactor, it is important to obtain the knowledge of rheological properties of all materials at different stages including raw lignocellulose material, pretreated material, hydrolysate, and fermentation slurry [81]. The rheological knowledge is the basis for the design of large-scale industrial reactor using computational dynamic fluid (CFD) method. Based on the information about the solid lignocellulosic materials-water system, CFD model can be developed and the mixing time and mechanical mixing energy can be calculated in order to assess the mixing efficiency [64, 82–85]. Meanwhile, the mock-up experiments should be conducted. Then, the simulation results should be compared with the mock-up experiments to verify the CFD model. If they were not in agreement very well, the former two steps should be carefully checked. If the validity of the CFD model was demonstrated by the mock-up results, then the third step, optimization of the configuration of pretreatment reactors by CFD simulation, can be conducted. The diameter, the pitch, the thickness of the impeller and the baffle design as well as the arrangement of the measure meters should all be considered. All these results will provide a strong design basis for the practical pretreatment reactor processing and will also guide for the reactor scaling-up.

Several pretreatment methods have been demonstrated at high solid loading in the pilot-scale cellulosic ethanol production plants, including dilute acid and steam explosion. However, the details in the pretreatment reactors are still limited [2, 12]. Recently, two industrial-scale ethanol production plants were declared to be open in Europe and USA, respectively (www.betarenewables.com; www.dsm.com), and several more industrial cellulosic ethanol plants are under construction [86]. It is expected that the forthcoming operation data could be released, which are very valuable and helpful for the future design and optimization of industrial pretreatment reactors for high solid loading.

Acknowledgments This research was supported by the National Basic Research Program of China (2011CB707406/2013CB733902), the Natural Science Foundation of China (21306048), the National High-Tech Program of China (2012AA022301/2014AA021901), and the Fundamental Research Funds for the Central Universities of China (WF1214025).

References

1. Galbe M, Zacchi G (2007) Pretreatment of lignocellulosic materials for efficient bioethanol production. Adv Biochem Engin Biotechnol 108:41–65
2. Jorgensen H, Kristensen JB, Felby C (2007) Enzymatic conversion of lignocellulose into fermentable sugars: challenges and opportunities. Biofuels Bioprod Bioref 1:119–134
3. Chandra RP, Bura R, Mabee WE, Berlin A, Pan X, Saddler JN (2007) Substrate pretreatment: the key to effective enzymatic hydrolysis of lignocellulosics? Adv Biochem Engin Biotechnol 108:67–93
4. Hendriks ATWM, Zeeman G (2009) Pretreatments to enhance the digestibility of lignocellulosic biomass. Bioresour Technol 100:10–18
5. Alvira P, Tomas-Pejo E, Ballesteros M, Negro MJ (2010) Pretreatment technologies for an efficient bioethanol production process based on enzymatic hydrolysis: a review. Bioresour Technol 101:4851–4861
6. Agbor VB, Cicek N, Sparling R, Berlin A, Levin DB (2011) Biomass pretreatment: fundamentals toward application. Biotechnol Adv 29:675–685
7. Galbe M, Zacchi G (2012) Pretreatment: the key to efficient utilization of lignocellulosic materials. Biomass Bioenergy 46:70–78
8. Yang B, Wyman CE (2008) Pretreatment: the key to unlocking low-cost cellulosic ethanol. Biofuels Bioprod Bioref 2:26–40
9. Mosier N, Wyman C, Dale B, Elander R, Lee YY, Holtzapple M, Ladisch M (2005) Features of promising technologies for pretreatment of lignocellulosic biomass. Bioresour Technol 96:673–686
10. Kumar P, Barrett DM, Delwiche MJ, Stroeve P (2009) Methods for pretreatment of lignocellulosic biomass for efficient hydrolysis and biofuel production. Ind Eng Chem Res 48:3713–3729
11. Sathitsuksanoh N, George A, Zhang Y-HP (2013) New lignocellulose pretreatments by using cellulose solvents: a review. J Chem Technol Biotechnol 88:169–180
12. Modenbach AA, Nokes SE (2012) The use of high-solids loadings in biomass pretreatment-A review. Biotechnol Bioeng 109:1430–1442
13. Zhang J, Wang XS, Chu DQ, He YQ, Bao J (2011) Dry pretreatment of lignocellulose with extremely low steam and water usage for bioethanol production. Bioresour Technol 102:4480–4488

14. He YQ, Zhang LP, Zhang J, Bao J (2014) Helically agitated mixing in dry dilute acid pretreatment enhances the bioconversion of corn stover into ethanol. Biotechnol Biofuels 7:1
15. Fan ZL, South C, Lyford K, Munsie J, van Walsum P, Lynd LR (2003) Conversion of paper sludge to ethanol in a semicontinuous solids-fed reactor. Bioprocess Biosyst Eng 26:93–101
16. Wingren A, Galbe M, Zacchi G (2003) Techno-economic evaluation of producing ethanol from softwood: comparison of SSF and SHF and identification of bottlenecks. Biotechnol Prog 19:1109–1117
17. Jorgensen H, Vibe-Pedersen J, Larsen J, Felby C (2007) Liquefaction of lignocellulose at high-solids concentrations. Biotechnol Bioeng 96:862–870
18. Palmowski L, Muller J (1999) Influence of the size reduction of organic waste on their anaerobic digestion. In: II International symposium on anaerobic digestion of solid waste. Barcelona, pp 137–144, 15–17 June
19. Sun Y, Cheng JY (2002) Hydrolysis of lignocellulosic materials for ethanol production: a review. Bioresour Technol 83:1–11
20. Zhu JY (2011) Physical pretreatment—woody biomass size reduction-for forest biorefinery. In: sustainable production of fuels, chemicals, and fibers from forest biomass; ACS Symposium Series, American Chemical Society, Washington, DC
21. Silva GGD, Couturier M, Berrin JG, Buleon A, Rouau X (2012) Effects of grinding processes on enzymatic degradation of wheat straw. Bioresour Technol 103:192–200
22. Schell DJ, Harwood C (1994) Milling of lignocellulosic biomass: results of pilot-scale testing. Appl Biochem Biotechnol 45(46):159–168
23. Mani S, Tabil LG, Sokhansanj S (2004) Grinding performance and physical properties of wheat and barley straws, corn stover and switchgrass. Biomass Bioenergy 27:339–352
24. Adapa P, Tabil L, Schoenau G (2011) Grinding performance and physical properties of non-treated and steam exploded barley, canola, oat and wheat straw. Biomass Bioenergy 35: 549–561
25. Kratky L, Jirout T (2011) Biomass size reduction machines for enhancing biogas production. Chem Eng Technol 34:391–399
26. Hideno A, Inoue H, Tsukahara K, Fujimoto S, Minowa T, Inoue S, Endo T, Sawayama S (2009) Wet disk milling pretreatment without sulfuric acid for enzymatic hydrolysis of rice straw. Bioresour Technol 100:2706–2711
27. Zhang J, Wang MY, Gao MT, Fang X, Yano S, Qin SL, Xia RR (2013) Efficient acetone-butanol-ethanol production from corncob with a new pretreatment technology-wet disk milling. Bioenergy Res 6:35–43
28. Silva ASD, Inoue H, Endo T, Yano S, Bon EPS (2010) Milling pretreatment of sugarcane bagasse and straw for enzymatic hydrolysis and ethanol fermentation. Bioresour Technol 101:7402–7409
29. Peng HD, Li HQ, Luo H, Xu J (2013) A novel combined pretreatment of ball milling and microwave irradiation for enhancing enzymatic hydrolysis of microcrystalline cellulose. Bioresour Technol 130:81–87
30. Teramoto Y, Tanaka N, Lee SH, Endo T (2008) Pretreatment of eucalyptus wood chips for enzymatic saccharification using combined sulfuric acid-free ethanol cooking and ball milling. Biotechnol Bioeng 99:75–85
31. Inoue H, Yano S, Endo T, Sakaki T, Sawayama S (2008) Combining hot-compressed water and ball milling pretreatments to improve the efficiency of the enzymatic hydrolysis of eucalyptus. Biotechnol Biofuels 1:2
32. Zhu JY, Zhuang XS (2012) Conceptual net energy output for biofuel production from lignocellulosic biomass through biorefining. Prog Energy Combust Sci 38:583–598
33. Zhu JY, Wang GS, Pan XJ, Gleisner R (2009) Specific surface to evaluate the efficiencies of milling and pretreatment of wood for enzymatic saccharification. Chem Eng Sci. 64:474–485
34. Li HJ, Ye CL, Liu K, Gu HQ, Du WT, Bao J (2014) Analysis of particle size reduction on overall surface area and enzymatic hydrolysis yield of corn stover. Bioprocess Biosys Eng. doi:10.1007/s00449-014-1253-y

35. Zhu W, Zhu JY, Gleisner R, Pan XJ (2010) On energy consumption for size-reduction and yield from subsequent enzymatic saccharification of pretreated lodgepole pine. Bioresour Technol 101:2782–2792
36. Chen XW, Tao L, Shekiro J, Mohaghaghi A, Decker S, Wang W, Smith H, Park S, Himmel ME, Tucher M (2012) Improved ethanol yield and reduced minimum ethanol selling price (MESP) by modifying low severity dilute acid pretreatment with deacetylation and mechanical refining: 1) experimental. Biotechnol Biofuels 5:60
37. Tao L, Chen XW, Aden A, Kuhn E, Himmel ME, Tucker M, Franden MAA, Zhang M, Johnson DK, Dowe N, Elander RT (2012) Improved ethanol yield and reduced minimum ethanol selling price (MESP) by modifying low severity dilute acid pretreatment with deacetylation and mechanical refining: 2) techno-economic analysis. Biotechnol Biofuels 5:69
38. Lin ZX, Huang H, Zhang HM, Zhang L, Yan LS, Chen JW (2010) Ball milling pretreatment of corn stover for enhancing the efficiency of enzymatic hydrolysis. Appl Biochem Bioethanol 162:1872–1880
39. Luo W, Wang J, Liu XB, Li HG, Pan HP, Gu QY, Yu XB (2013) A facile and efficient pretreatment of corncob for bioproduction of butanol. Bioresour Technol 140:86–89
40. Kim HJ, Lee S, Kim J, Mitchell RJ, Lee JH (2013) Environmentally friendly pretreatment of plant biomass by planetary and attrition milling. Bioresour Technol 144:50–56
41. Trater AM, Alavi S, Rizvi SSH (2005) Use of non-invasive X-ray microtomography for characterizing microstructure of extruded biopolymer foams. Food Res Int 38:709–719
42. Lee SH, Teramoto Y, Endo T (2009) Enzymatic saccharification of woody biomass micro/nanofibrillated by continuous extrusion process I—effect of additives with cellulose affinity. Bioresour Technol 100:275–279
43. Zhang SJ, Xu YX, Hanna MA (2012) Pretreatment of corn stover with twin-screw extrusion followed by enzymatic saccharification. Appl Biochem Biotechnol 166:458–469
44. Liu C, van der Heide E, Wang HS, Li B, Yu G, Mu XD (2013) Alkaline twin-screw extrusion pretreatment for fermentable sugar production. Biotechnol Biofuels 6:97
45. Karunanithy C, Muthukumarappan K (2010) Influence of extruder temperature and screw speed on pretreatment of corn stover while varying enzymes and their ratios. Appl Biochem Biotechnol 162:264–279
46. Choi CH, Oh KK (2012) Application of a continuous twin-driven process for dilute acid pretreatment of rape straw. Bioresour Technol 110:349–354
47. Karunanithy C, Muthukumarappan K (2011) Optimization of switchgrass and extruder parameters for enzymatic hydrolysis using response surface methodology. Ind Crop Prod 33:188–199
48. Karunanithy C, Muthukumarappan K, Gibbons WR (2012) Extrusion pretreatment of pine wood chips. Appl Biochem Biotechnol 167:81–99
49. Yoo J, Alavi S, Vadlani P, Behnke KC (2012) Soybean hulls pretreated using thermo-mechanical extrusion-Hydrolysis efficiency, fermentation inhibitors, and ethanol yield. Appl Biochem Biotechnol 166:576–589
50. Chen WH, Xu YY, Hwang WS, Wang JB (2011) Pretreatment of rice straw using an extrusion/extraction process at bench-scale for producing cellulosic ethanol. Bioresour Technol 102:10451–10458
51. Chen JW, Zhang WG, Zhang HM, Zhang QX, Huang H (2014) Screw extrude steam explosion: a promising pretreatment of corn stover to enhance enzymatic hydrolysis. Bioresour Technol 161:230–235
52. Zhang YD, Hou TA, Li B, Liu C, Mu XD, Wang HS (2014) Acetone-butanol-ethanol production from corn stover pretreated by alkaline twin-screw extrusion pretreatment. Bioprocess Biosyst Eng 37:913–921
53. Duque A, Manzanares P, Ballesteros I, Negro MJ, Oliva JM, Gonzalez A, Ballesteros M (2014) Sugar production from barley straw biomass pretreated by combined alkali and enzymatic extrusion. Bioresour Technol 158:262–268

54. Humbird D, Davis R, Tao L, Kinchin C, Hsu D, Aden A, Schoen P, Lukas J, Olthof B, Worley M, Sexton D, Dudgeon D (2011) Process design and economics for biochemical conversion of lignocellulosic biomass to ethanol. National Renewable Energy Laboratory, Golden CO
55. Lamsal B, Yoo J, Brijwani K, Alavi S (2010) Extrusion as a thermo-mechanical pre-treatment for lignocellulosic ethanol. Biomass Bioenergy 34:1703–1710
56. Senturk-Ozer S, Gevgilili H, Kalyon DM (2011) Biomass pretreatment strategies via control of rheological behavior of biomass suspensions and reactive twin screw extrusion processing. Bioresour Technol 102:9068–9075
57. Yoo J, Alavi S, Vadlani P, Amanor-Boadu V (2011) Thermo-mechanical extrusion pretreatment for conversion of soybean hulls to fermentable sugars. Bioresour Technol 102:7583–7590
58. Eckard AD, Muthukumarappan K, Gibbons W (2012) Pretreatment of extruded corn stover with polyethylene glycol to enhance enzymatic hydrolysis: optimization, kinetics, and mechanism of action. Bioenergy Res 5:424–438
59. Msaon WH (1926) Process and apparatus for disintegration of wood and the like. US Patent: 1578609
60. Palmaqvist E, Hahn-Hagerdal B, Galbe M, Larsson M, Stenberg K, Szengyel Z, Tengborg C, Zacchi G (1996) Design and operation of a bench-scale process development unit for the production of ethanol from lignocellulosics. Bioresour Technol 58:171–179
61. Alizadeh H, Teymouri F, Gilbert TI, Dale BE (2005) Pretreatment of switchgrass by ammonia fiber explosion (AFEX). Appl Biochem Biotechnol 121–124:1133–1141
62. Kim HT, Lee YY (2005) Pretreatment and fractionation of corn stover by ammonia recycle percolation process. Bioreousr Technol 96:2007–2013
63. Narayanaswamy N, Faik A, Goetz DJ, Gu TY (2011) Supercritical carbon dioxide pretreatment of corn stover and switchgrass for lignocellulosic ethanol production. Bioresour Technol 202:6995–7000
64. Hamelinck CN, van Hooijdonk G, Faaij APC (2005) Ethanol from lignocellulosic biomass: techno-economic performance in short-, middle- and long-term. Biomass Bioenergy 28: 384–410
65. Linde M, Jakobsson EL, Galbe M, Zacchi G (2008) Steam pretreatment of dilute H_2SO_4-impregnated wheat straw and SSF with low yeast and enzyme loadings for bioethanol production. Biomass Bioenergy 32:326–332
66. Carrasco C, Baudel HM, Sendelius J, Modig T, Roslander C, Galbe M, Hahn-Hagerdal B, Zacchi G, Liden G (2008) SO_2-catalyzed steam pretreatment and fermentation of enzymatically hydrolyzed sugarcane bagasse. Enzyme Microb Technol 46:64–73
67. Jin MJ, Gunawan C, Balan V, Dale BE (2012) Consolidated bioprocessing (CBP) of AFEX^{TM}-pretreated corn stover for ethanol production using Clostridium phytofermentans at a high solids loading. Biotechnol Bioeng 109:1929–1936
68. Bals B, Dale B, Balan V (2006) Enzymatic hydrolysis of distiller's dry grain and solubles (DDGS) using ammonia fiber expansion pretreatment. Energy Fules 20:2732–2736
69. Balan V, Sousa LDC, Chundawat SPS, Marshall D, Sharma LN, Chambliss CK, Dale BE (2009) Enzymatic digestibility and pretreatment degradation products of AFEX-treated hardwoods (Populus nigra). Biotechnol Progr 25:265–275
70. Murnen HK, Balan V, Chundawat SPS, Bals B, Sousa LDC, Dale BE (2007) Optimization of ammonia fiber expansion (AFEX) pretreatment and enzymatic hydrolysis of miscanthus x giganteus to fermentable sugars. Biotechnol Progr 23:846–850
71. Laser M, Jin HM, Jayawardhana K, Lynd LR (2009) Coproduction of ethanol and power from switchgrass. Biofuels Bioprod Bioref 3:195–218
72. Zhong C, Lau MW, Balan V, Dale BE, Yuan YJ (2009) Optimization of enzymatic hydrolysis and ethanol fermentation from AFEX-treated rice straw. Appl Microbiol Biotechnol 84: 667–676
73. Li BZ, Balan V, Yuan YJ, Dale BE (2010) Process optimization to convert forage and sweet sorghum bagasse to ethanol based on ammonia fiber expansion (AFEX) pretreatment. Bioresour Technol 101:1285–1292

74. Kumar R, Mago G, Balan V, Wyman CE (2009) Physical and chemical characterization of corn stover and poplar solids resulting from leading pretreatment technologies. Bioresour Technol 100:3948–3962
75. Zhang J, Chu DQ, Huang J, Yu ZC, Dai GC, Bao J (2010) Simultaneous saccharification and ethanol fermentation at high corn stover solids loading in a helical stirring bioreactor. Biotechnol Bioeng 105:718–728
76. Zhang LP, Zhang J, Li CH, Bao J (2014) Rheological characterization and CFD modeling of corn stover-water mixing system at high solids loading for dilute acid pretreatment. Biochem Eng J 90:324–332
77. He YQ, Zhang J, Bao J (2014) Dry dilute acid pretreatment by co-currently feeding of corn stover feedstock and dilute acid solution without impregnation. Bioresour Technol 158:360–364
78. Viamajala K, Mcmillan JD, Schell DJ, Elander RT (2009) Rheology of corn stover slurries at high solids concentrations: effects of saccharification and particle size. Bioresour Technol 100:925–934
79. Kemblowski Z, Kristiansen B (1986) Rheology of fermentation liquids. Biotechnol Bioeng 28:1474–1483
80. Lavenson DM, Tozzi EJ, McCarthy MJ, Powell RL (2011) Yield stress of pretreated corn stover suspensions using magnetic resonance imaging. Biotechnol Bioeng 108:2312–2319
81. Winkler MA (1990) Problems in fermenter design and operation. In: Winkler MA (ed) Chemical engineering problems in biotechnology. Elsevier Science Publishers LTD., London, pp 215–337
82. Renzo AD, Maio DFP (2007) Homogeneous and bubbling fluidization regimes in DEM-CFD simulations: hydrodynamic stability of gas and liquid beds. Chem Eng Sci 62:116–130
83. Habla F, Marschall H, Hinrichsen O, Dietsche L, Jasak H, Favero JL (2011) Numerical simulation of viscoelastic two-phase flows using open FOAM. Chem Eng Sci 66:5487–5496
84. Giuseppina M, Michal M, Milan J, Franco M (2005) CFD simulations and experimental validation of homogenization curves and mixing time in stirred Newtonian and pseudoplastic liquids. Chem Eng Sci 60:2427–2437
85. Yu L, Ma J, Chen S (2011) Numerical simulation of mechanical mixing in high solid anaerobic digester. Bioresour Technol 102:1012–1018
86. Brown TR, Brown RC (2013) A review of cellulosic biofuel commercial-scale projects in the United States. Biofuels Bioprod Bioref 7:235–245

Adv Biochem Eng Biotechnol (2016) 152: 91–136
DOI 10.1007/10_2015_326

Published Online: 16 May 2015

Perturbation Experiments: Approaches for Metabolic Pathway Analysis in Bioreactors

Michael Weiner, Julia Tröndle, Christoph Albermann, Georg A. Sprenger and Dirk Weuster-Botz

Abstract In the last decades, targeted metabolic engineering of microbial cells has become one of the major tools in bioprocess design and optimization. For successful application, a detailed knowledge is necessary about the relevant metabolic pathways and their regulation inside the cells. Since in vitro experiments cannot display process conditions and behavior properly, process data about the cells' metabolic state have to be collected in vivo. For this purpose, special techniques and methods are necessary. Therefore, most techniques enabling in vivo characterization of metabolic pathways rely on perturbation experiments, which can be divided into dynamic and steady-state approaches. To avoid any process disturbance, approaches which enable perturbation of cell metabolism in parallel to the continuing production process are reasonable. Furthermore, the fast dynamics of microbial production processes amplifies the need of parallelized data generation. These points motivate the development of a parallelized approach for multiple metabolic perturbation experiments outside the operating production reactor. An appropriate approach for in vivo characterization of metabolic pathways is presented and applied exemplarily to a microbial L-phenylalanine production process on a 15 L-scale.

Keywords Perturbation experiments · Steady-state experiments · Metabolic flux analysis · Constraint-based approaches · Metabolome quantification · L-phenylalanine · *Escherichia coli* · Glycerol

Abbreviations

μLC	Microcapillary liquid chromatography
^{13}C-MFA	Labeling experiments for metabolic flux analysis

M. Weiner · J. Tröndle · D. Weuster-Botz (✉)
Lehrstuhl für Bioverfahrenstechnik, Technische Universität München, Garching, Germany
e-mail: d.weuster-botz@lrz.tum.de

C. Albermann · G.A. Sprenger
Institut für Mikrobiologie, Universität Stuttgart, Stuttgart, Germany

13DPG	1,3-bisphospho-D-glycerate
2PG	2-phospho-D-glycerate
3DHS	3-dehydroshikimate
3PG	3-phospho-D-glycerate
AcCoA	Acetyl coenzyme A
AKG	α-ketoglutarate
CDW	Cell dry weight
CE	Capillary electrophoresis
CIT	Citrate
CER	Carbon dioxide evolution rate
COBRA	Constraint-based reconstruction and analysis
DAHP	3-deoxy-D-arabino-heptulosonate-7-phosphate
DHAP	Dihydroxyacetone phosphate
E. coli	*Escherichia coli*
E4P	Erythrose-4-phosphate
ESI	Electrospray ionization
F6P	Fructose-6-phosphate
FBA	Flux balance analysis
FBP	Fructose-1,6-bisphosphate
FUM	Fumarate
FVA	Flux variability analysis
G6P	Glucose-6-phosphate
GAP	Glyceraldehyde-3-phosphate
GC	Gas chromatography
*g*FBA	Geometric flux balance analysis
GLYC	Glycerol
ICIT	Isocitrate
IDMS	Isotope dilution mass spectrometry
IPTG	Isopropyl β-D-1-thiogalactopyranoside
LC	Liquid chromatography
*ll*FVA	Loopless FVA
LP	Linear programming
L-phe	L-phenylalanine
L-trp	L-tryptophan
L-tyr	L-tyrosine
m/z	Mass to charge ratio
MAL	Malate
MALDI	Matrix-assisted laser desorption ionization
MCA	Metabolic control analysis
MFA	Metabolic flux analysis
MS	Mass spectrometry
NMR	Nuclear magnetic resonance
OAA	Oxaloacetate
OUR	Oxygen uptake rate

PEP	Phosphoenolpyruvate
PPHN	Prephenate
PYR	Pyruvate
R5P	Ribose-5-phosphate
Ru5P	Ribulose-5-phosphate
S7P	Sedoheptulose-7-phosphate
SHIK	Shikimate
S3P	Shikimate-3-phosphate
SUC	Succinate
SuccCoA	Succinyl coenzyme A
*t*FVA	Thermodynamically constrained FVA
TOF	Time of flight
UHPLC	Ultrahigh-performance liquid chromatography
U-^{13}C	Uniform ^{13}C labeled
X5P	Xylulose-5-phosphate

Contents

1 Introduction 94
2 Perturbation Experiments 95
2.1 Dynamic Perturbation Experiments 95
2.2 Steady-State Perturbation Experiments 96
3 Metabolic Flux Analysis (MFA) 98
3.1 ^{13}C-based Metabolic Flux Analysis 98
3.2 Constraint-Based Approaches 99
4 Metabolomics 102
4.1 Sampling 104
4.2 Quenching 105
4.3 Extraction 106
4.4 Quantification 107
4.5 Analytics 108
5 Perturbation Experiments During Microbial L-phenylalanine Production 109
5.1 Industrial Relevance of Microbial L-phenylalanine Production 109
5.2 L-phenylalanine Production Process 110
5.3 Parallelized Batch Steady-State Perturbation Experiments During L-phenylalanine Production 112
5.4 Parallelized Fed-Batch Steady-State Perturbation Experiments During L-phenylalanine Production 116
6 Conclusions 128
References 129

1 Introduction

Biotechnological processes for the production of fine chemicals are of special interest because of their high selectivity and specificity while at the same time operating under mild conditions. The two major starting points for process optimization are seen in the adjustment of process parameters and in genetic modifications of the applied production strain. The latter used to be performed by random trial-and-error genetic modifications, which is nowadays replaced by increasingly targeted metabolic engineering [1]. For this, a detailed knowledge about the relevant metabolic pathways is necessary.

There are numerous advantages of using whole cells during production processes rather than isolated enzymes, especially if more than one enzymatic step is necessary for the synthesis aimed at. They include the integrated recovery of cofactors and the avoidance of laborious purification of enzymes. A possible drawback is the integration of the production pathway in the complex cellular metabolism, because this may lead to limitations during production due to regulations present in the cell [2]. For a successful rational optimization, it is thus necessary to take a broad view on metabolism while at the same time considering the conditions that are present during the process. That means that data about the cells' metabolic state in the process have to be collected in vivo, which requires special techniques and methods.

Due to the high complexity of metabolism even in the simplest cells, mathematical models are indispensable to gain knowledge and understanding of them. Such models have been evolved for several decades now and are available for numerous organisms. The long-term aim is models that are able to describe the complete organisms in mechanistic detail, including all regulations spreading over the hierarchical layers of metabolism [3]. Today, it is possible either to describe a selected section of metabolism mechanistically, i.e., including enzyme kinetics, or to represent the metabolism on genome scale in a constraint-based model. Both types of models can help to interpret data gained about metabolism and to find new targets for strain optimization. Therefore, they are particularly capable of revealing non-obvious targets and possible secondary effects of genetic modifications [4]. In the future, it might be possible to design production strains solely on the basis of theoretical estimations, but today, experimental data generation and verification are crucial to check the predictions and to further refine the models.

Here, we highlight several possibilities to gain information about the metabolic state of microorganisms. We give an overview of various methods to perform perturbation experiments, which usually are the basis for detailed metabolic analyses and can be divided into dynamic and steady-state experiments. This is complemented by a description of the usual possibilities for metabolic flux analysis. They rely on theoretical estimations and comprise ^{13}C-labeling experiments as well as constraint-based methods. Another topic discussed in this context is the quantification of intracellular metabolites, which requires several special techniques. Moreover, we present an improved method for metabolic data generation in parallel to the operating cultivation with minimal disturbance of the process and give an example for the characterization of an industrially relevant L-phenylalanine production process with recombinant *Escherichia coli*.

2 Perturbation Experiments

One central aspect of metabolic engineering is the identification of parameters for rational and targeted modification of cellular metabolism [5]. Therefore, detailed knowledge about the relevant enzyme properties and the pathways of interest is essential. Perturbation studies are experimental approaches which can provide the information in vivo. Since in vivo analysis is a very challenging task, a wide range of methods and techniques is available and was used in the last years. Perturbation experiments are commonly divided into dynamic and stationary studies. These approaches will be presented separately in the next chapters.

2.1 Dynamic Perturbation Experiments

Dynamic perturbation experiments, also referred to as stimulus–response experiments or pulse experiments, are a technique especially used for identification of in vivo kinetics [6, 7].

For these experiments, a continuous process serves as a long-time metabolic steady state. Perturbation of this state is mostly performed by a rapid change of substrate availability. Therefore, either a temporary switch from the continuous carbon source to another carbon source is applied [8], or substrate availability of the continuous source is drastically changed (e.g., substrate pulse) [8–13]. Beside substrate availability, perturbation of metabolism can also be realized by rapid changes of essential process parameters such as oxygen supply [10]. Using these perturbation techniques enables a metabolic shift from the steady state, and a time-dependent change of the intracellular metabolome is obtained [7]. This dynamic cellular response is characterized by collecting a high number of samples within seconds after the metabolic perturbation is applied. For correct quantification of intracellular metabolites, high-throughput methods of rapid sampling, quenching, subsequent extraction, and quantification techniques are necessary (see Sect. 4). In many cases, dynamic perturbation experiments are performed by perturbing the steady state directly in the laboratory-scale stirred-tank bioreactor [9–12]. Unfortunately, starting up and running a laboratory-scale continuous cultivation until a metabolic steady state is reached can be a time-consuming procedure. Further, it must be considered that the process may take a long time to equilibrate to the former steady state after a perturbation has been applied. In worst case, only one perturbation experiment can be performed during a steady-state culture [14].

Because of these reasons, instrument-based techniques such as the BioScope reactor [15, 16] are available, which enable dynamic perturbation experiments outside the steady-state bioreactor. As a consequence, this small-scale analytical reactor enables multiple perturbations to be applied sequentially during the same steady-state process.

The BioScope is the most intensively described instrument-based technique for perturbation experiments outside the continuous process. This reactor, initially described by [15], is based on a plug-flow reactor design with two hemispherical channels separated by a silicone membrane that allows diffusion of O_2 and CO_2 [16]. A small channel volume of 3.46 mL in combination with a total channel length of 6.51 m results in a high surface to volume ratio. For the second generation of the BioScope reactor, an overall O_2 mass transfer coefficient of $1.8\ e^{-5}\ m\ s^{-1} \pm 5\ \%$ is described [16]. Perturbation experiments with the BioScope reactor are initiated by a constant flow of culture broth from the steady-state bioreactor into the BioScope inlet port. Perturbation is applied by adding a constant flow of perturbation agent at the beginning of the plug-flow reactor. Multiple sampling ports are located at different distances of the reaction channel to enable a good time resolution of perturbation dynamics. Due to the constant flow rate of the disturbed culture broth through the plug-flow reactor, the distance between the point of mixing and the point of sampling determines the time during which the sampled cells are exposed to the perturbation [15].

2.2 *Steady-State Perturbation Experiments*

An alternative to dynamic perturbations are steady-state perturbations. Within these experiments, multiple metabolic steady states are set up to characterize the cellular metabolism. Comparable to dynamic experiments, a continuous process is mostly used to realize an initial metabolic steady state [17, 18]. Beside chemostat experiments, fed-batch cultivations with a growth-rate-controlled feeding strategy were shown to be suitable for steady-state perturbation experiments [19]. Perturbation of the initial state is performed by an enduring change in environmental conditions. For this purpose, different substrates [19], changes in substrate supply rate [18], or process parameters (e.g., variation of pH [17]) are applied. In contrast to dynamic experiments, where the initial dynamic response to the perturbation is the central focus, the new metabolic steady state is of special interest during steady-state perturbation experiments. For characterization of this novel metabolic steady state, changes in extracellular fluxome (e.g., substrate uptake, production rate and respiration rates), intracellular fluxome, and in intracellular metabolome are studied [17–19].

The advantage of steady-state perturbation experiments compared to dynamic experiments is generally seen in the reduced number of metabolic states that have to be analyzed, whereas for characterization of dynamic behavior, much more samples and precise analytical devices are necessary for a sufficient time resolution [19]. Furthermore, a new metabolic steady state can be achieved within minutes in stationary perturbation experiments. Within these short-term experiments, a constant enzyme level is assumed [19]. This point makes steady-state perturbation experiments suitable for data generation with constant enzyme levels, which are used in control theorems such as 'metabolic control analysis (MCA)'. Within MCA, fluxes and metabolite levels of multiple steady states are correlated for estimation of control coefficients.

In the last years, several experiments were described in which different steady states were realized subsequently in bioprocesses [17, 18]. Since re-equilibration to the initial steady state after perturbation of metabolism can take a long time or may not be achieved at all for fast changing cultivations with recombinant strains, perturbations within the same stirred-tank bioreactor may influence process behavior [14]. Because of this, comparable to dynamic perturbation experiments, instrumental techniques and methods, which allow perturbation of metabolism outside of the steady-state bioprocess, are highly advantageous for steady-state perturbation experiments. In the last years, instrumental techniques such as the Biocurve reactor [14] and methods such as 'Rapid Media Transition' [19] have been published.

The concept of the Biocurve reactor is similar to the BioScope reactor (see Sect. 2.1). For perturbation experiments, culture broth is pumped from the steady-state process into the Biocurve reactor. Two hemispherical flow channels separated by a silicone membrane are used for broth transport within the reactor. For the Biocurve reactor, an O_2 mass transfer coefficient of $1.55\ e^{-5} \pm 0.17\ m\ s^{-1}$ is described [14]. In contrast to the BioScope reactor, a recycle flow is used for ideal broth mixing in the Biocurve reactor. As a result of these high linear velocities and the resulting broth mixing, application of Biocurve reactor for steady-state perturbation experiments is possible [14].

The experimental method of 'Rapid Media Transition' also enables performance of steady-state perturbation of metabolism in parallel to the operating bioprocess [19]. For this purpose, cells are separated from the cultivation process by rapid centrifugation. Supernatant is discarded and cells are transferred with fresh media into a separate laboratory-scale stirred-tank reactor. In this analytical scale, perturbation of metabolism can be applied by modification of environmental conditions (e.g., different substrate sources). In contrast to the approaches of Biocurve and BioScope reactor, no specially designed bioreactor is necessary. 'Rapid Media Transition' experiments can be performed with standardized, commercially available laboratory-scale stirred-tank bioreactor systems. This point also implies that standardized and well-established online sensors for pH, dissolved oxygen, and off-gas components such as O_2 and CO_2 are applicable. Further, pH control and substrate addition by peristaltic milliliter-scale pumps can be achieved easily.

3 Metabolic Flux Analysis (MFA)

In contrast to metabolome, proteome, and transcriptome data, intracellular fluxes cannot be measured directly [20]. For that reason, model-based computational analysis supported by measurable quantities such as extracellular fluxes or intracellular metabolites is necessary for the estimation of intracellular fluxes.

Estimation of intracellular fluxes is described by the broad field of metabolic flux analysis (MFA). In general, MFA includes the three different fields of stoichiometric metabolite balancing, constraint-based approaches, and ^{13}C-based metabolic flux analysis.

Stoichiometric metabolite balancing is based solely on the measurement of extracellular fluxes [5]. For mathematical feasibility, a network-dependent amount of fluxes has to be measurable. As a result, a strong simplification of the metabolic network is necessary.

In contrast to this approach, more complex networks and/or genome-scale models can be considered with ^{13}C-based metabolic flux analysis and constraint-based approaches. For that reason, these two approaches are summarized subsequently.

3.1 ^{13}C-based Metabolic Flux Analysis

In the last decade, intensive research work on ^{13}C-based metabolic flux analysis (^{13}C-MFA) has been performed. In combination with significant improvement of analytical methods, computational capabilities, and software tools [21–23], the methodical spectrum of ^{13}C-MFA was significantly extended [20, 24–35].

All approaches of ^{13}C-MFA have in common that ^{13}C-labeled substrates are fed to a population and the resulting labeling pattern of relevant metabolic intermediates is quantified by nuclear magnetic resonance (NMR) or mass spectrometry (MS) for the determination of intracellular fluxes.

Classical ^{13}C-MFA is performed under metabolic and isotopic stationary conditions [20, 29, 32], which can be realized, for example, by chemostat cultures. In classical ^{13}C-MFA, most frequently ^{13}C-patterns of protein-bound amino acids are quantified. These amino acids are carbon backbones of key intermediates of central metabolism, which enable determination of central metabolic fluxes during steady state. Estimation of fluxes is realized by iterative data fitting, whereby the differences between measured and simulated labeling patterns are minimized [29]. Within this parameter-fitting procedure, mathematical models of varying complexity are used [20]. The advantage of this tracer method is seen in the high protein stability and abundance. As a negative consequence of this, it takes a long labeling time before the protein is in isotopic steady state, which increases experimental costs. To obtain more information or information beyond fluxes of central metabolism, the labeling pattern of other intracellular metabolites has to be measured. Due to their low intracellular concentration, high turnover rates, and their diverse stability, this point is very challenging but nowadays realizable (see Sect. 4). Since the protein pool is connected to metabolism in both directions (synthesis/degradation), also stationary labeling of intracellular metabolite pools requires long labeling time [35]. Due to this, isotopic stationary ^{13}C-MFA is not applicable for industrial production processes with fast changing conditions, non-growing cells, or high-throughput experiments [31, 36].

Because of these methodical limitations of isotopic stationary ^{13}C-MFA, isotopic non-stationary approaches (INST-^{13}C-MFA) have become of central interest in the last years [30, 31, 33–35, 37]. In INST-^{13}C-MFA experiments, a sudden change from a uniformly ^{12}C labeled substrate to a defined ^{13}C labeled mixture is performed under metabolic steady-state conditions. For analysis of the resulting dynamic changes in labeling pattern of intracellular metabolites, multiple samples are taken within a short timescale (seconds to minutes). For this purpose, rapid sampling and quenching systems for high sample throughput were established (see Sect. 4.1). Subsequent quantification of low metabolite concentrations and labeling pattern is realized by using newest state-of-the art analytical methods of liquid chromatography (LC)- and gas chromatography (GC)-MS (see Sect. 4.5). For computational description of these dynamic changes in labeling patterns, a high-dimensional system of differential equations has to be applied. Furthermore, numerical stiffness problems must be considered due to different timescales of label accumulation processes in the network [30]. Consequently, on the one hand, INST-^{13}C-MFA enables metabolic flux determination with a high resolution of only a few minutes [30]. On the other hand, INST-^{13}C-MFA is much more demanding than classical ^{13}C-MFA with respect to experimental, analytical, and computational effort.

3.2 Constraint-Based Approaches

Labeling experiments can provide kinetic models, which give detailed knowledge about cellular functions and thereby generate ultimate understanding about the cellular network. Unfortunately, these models are limited by the availability of the information which is needed to generate them [38]. Furthermore, experimental costs, analytical effort, and duration of data acquisition are high. As a methodological alternative, constraint-based approaches can be used to enable estimation of intracellular fluxes. These well-established approaches are described in several reviews (e.g., [39–42]) and were applied for recent studies of biochemical networks [1, 43, 44].

In general, constraint-based approaches do not primarily intend to determine a single solution set of intracellular fluxes. Within these approaches, an allowable solution space is generated by applying different inviolable rules of physico-chemical constraints. Information mostly used to define and reduce the allowable solution space includes stoichiometric constraints (mass balance), thermodynamical constraints (reaction reversibility and direction), and enzyme-specific constraints (appropriate v_{max}) [38].

Stoichiometric constraints are represented by the matrix equation **Sv** = **0**, where **S** (stoichiometric matrix; columns: reactions; rows: metabolites) describes all the reactions in the applied network. Fluxes for each reaction are displayed in the vector **v**. This mathematical description of the metabolism with each reaction represented by a linear equation is the basis for every constraint-based approach. **Sv** = **0** furthermore includes a steady-state assumption, which means that the total production of each metabolite equals its total consumption. In other words, no metabolite accumulates inside the cell. The solution space defined by this equation is highly under-determined and an indefinite number of solutions is possible.

Thermodynamical constraints further restrict the possible range of flux values. This comprises that an irreversible reaction has to be indicated by a flux value greater than or equal to zero. Enzymes that can catalyze both directions (reversible reactions) can have positive or negative flux values. A systematic illustration of thermodynamical constraints was reviewed in 2002 [45, 46].

Moreover, enzyme-specific constraints can be used to reduce solution space. Therefore, data from literature (e.g., v_{max}) or experimental data such as extracellular fluxes of substrate uptake, respiration rates, and production rates [47] can be used.

To apply accurate physicochemical constraints, extensive work was performed in the last decades to develop precise and accurate stoichiometric networks for constraint-based modeling. A historical review of 13 years of metabolic network improvement was exemplarily described for the bacterium *E. coli* [38]. Within this review, development of network models was demonstrated, starting with a simple network of 14 metabolic reactions in 1990 [48] and resulting in a genome-scale approach with 931 metabolic reactions (model *i*JR904; [49]). Since this review, further significant model improvement was achieved, advantaged by sequencing and annotation of the complete genome of *E. coli* K12 MG1655 and W3110 [50, 51]. Current genome-scale models of *E. coli* include 1387 (model *i*AF1260; [52]) to 1473 metabolic reactions (model *i*JO1366; [53]). Beside network reconstruction of *E. coli*, a significantly increasing number of genome-scale models are available for other organisms (e.g., *Saccharomyces cerevisiae* (model *i*FF708 [54]; model *i*IN800 [55]; model review in [56]); *Kluyveromyces lactis* (model *i*OD907 [57]); *Clostridium acetobutylicum* ATCC 824 (model *i*CAC490 [58]); and *Methanococcus maripaludis* S2 (model *i*MM518 [59]); *Gluconobacter oxydans* 621H (model *i*XW433 [60]). Furthermore, constraint-based approaches have become more public, since software tools, like the freely available MATLAB-based COBRA toolbox, are available [61] and have been further improved [62]. Furthermore, development and sharing of genome-scale models can be realized by a standardized model format (SBML: Systems Biology Markup Language [63]) and model-databases like BiGG [64].

The goal of constraint-based approaches is to predict which of the possible flux distributions are present in the cells under the conditions of interest [49]. For this purpose, classical linear optimization techniques are applied. This means that a model which is defined by constraints is combined with an objective function that is to be maximized or minimized. A flux distribution for optimal fulfillment of the objective function can be determined by linear optimization. Some commonly used objective functions include optimization of biomass formation, production of a desired metabolite, or energy formation (production of ATP) [65].

Constraint modeling techniques such as flux balance analysis (FBA) [41, 66, 67], flux variability analysis (FVA) [68, 69], and geometric flux balance analysis (gFBA) [70] enable the usage of linear optimization for the characterization of the constraint solution space. These exemplary techniques are summarized in the following sections.

3.2.1 Flux Balance Analysis (FBA)

In genome-scale metabolic network reconstructions, FBA is a widely used mathematical approach for analyzing the flow of metabolites through a metabolic network. For FBA, a constraint-based model and an objective function are applied for the determination of a single flux distribution by linear optimization. Therefore, FBA enables the prediction of the growth rate of an organism or the rate of production of a biotechnologically important metabolite [41].

A historical overview of early FBA approaches and its first applications was given by Schilling et al. [66] and Edwards and Palsson [67]. A more recent introduction to FBA methodology was presented by [41]. Therein the relevant steps of mathematical representation of metabolism, integration of constraints, and optimization criteria are described. Furthermore, applicability of FBA is highlighted for the growth-rate estimation on different media, simulation of knockout mutations, and yield estimation for essential cofactors such as ATP, NADH, or NADPH.

In general, methodical limitations of FBA are seen in the necessity of a present steady state, the non-consideration of regulatory effects [41], and its focus on single solution optimization.

3.2.2 Flux Variability Analysis (FVA)

FBA enables determination of one single solution set of intracellular fluxes which are defined by the flow of mass and energy. In the majority of cases, these flux distributions are not unique. For any given optimal flux distribution, there may exist several alternative optimal solutions that define a solution space fulfilling the same objective value (e.g., growth rate) and could represent biologically meaningful solutions [71]. These variances in flux solutions can be reasoned by the inherent redundancy of metabolism to enable intracellular fluxes on different evolutionary given pathways. For variance analysis, a FBA is initially performed for the quantification of the optimal solution. Starting from this solution, the range of variability for each flux is calculated through a series of linear programming (LP) problems. For this, the value of the original objective is fixed or tolerated within a definable percentage range. Each reaction in the network is then maximized and afterward minimized to calculate the range of metabolic flux that allows to meet the given scope of values for the objective [71]. To emphasize, this approach of variance analysis does not generate a list of single data sets which describe all optimal feasible solutions. Within this procedure, maximal and minimal fluxes are calculated which describe the range of all optimal solutions, separately for every reaction.

Variance analysis of constraint-based flux estimation also allows further interpretation of the used model and the constraint data set. High variances in flux estimation can, for example, be reasoned by two aspects: On the one side, variances in flux distribution can indicate alternative pathway routes. This information is essential for further pathway characterization and therefore of central interest in pathway analyses [44]. On the other side, high variances can also be reasoned by biologically irrelevant internal loops that satisfy any mathematical objective and therefore are not of interest in pathway analyses. These loops occur because constraint-based approaches ignore the imposition of the loop law (detailed in [69] and [46]). To consider the central aspect of loop law in constraint-based approaches, several approaches were recently presented. Within these methods, either further thermodynamic constraints [72–74] were used or methodology of loop identification and elimination was expanded (e.g., *ll*FVA/FBA [69] and *t*FVA [75]). A further alternative to eliminate loops in constraint-based flux estimation uses geometric modeling and is described in the following section.

3.2.3 Geometric Flux Balance Analysis (*g*FBA)

In constraint-based flux estimation approaches, biologically irrelevant internal flux loops can lead to a large number of flux distributions that satisfy all mathematical objectives. The relevance of *g*FBA can be seen in elimination of these irrelevant data sets. Within this approach, a geometric perspective is used to determine one single solution of flux distribution [70]. For this, finding of the smallest hull that contains all sets of optimal FBA solutions is posed as a set of multiple LP problems. Further, the assumption is made that the cell will minimize the total intracellular flux, which is necessary to fulfill the constraint objective. The idea of total flux minimization was first suggested by [76], who assumed that the total flux is proportional to the total enzyme concentration, which the cell is willing to minimize. This additional constraint is used to minimize the distance between the flux and the center of the convex hull. With each applied iteration, the allowable solution space is reduced by the algorithm. After a finite number of iterations, the bounds converge to a single solution, which is free of internal loops [70].

4 Metabolomics

The term metabolomics refers to the qualitative or quantitative analysis of all metabolites in and around cells at a certain time. The entity of all these metabolites (organic molecules <1500 Da) is termed metabolome. The metabolome contains particularly much information about the metabolic state of an organism because it is the metabolic level that reacts fastest to changes in environment [77]. Whereas the genome always remains constant and the proteome changes within many minutes or hours, metabolite levels may vary within a split of a second. While the metabolites comprise only 3 % of *E. coli* cell dry weight [78], this group contains very many substances with entirely diverging chemical properties whose concentrations vary around magnitudes. Additionally, they are consumed with high turnover rates. For all these reasons, it is an experimental challenge to qualify and quantify them correctly. Highly demanding techniques are necessary for sample collection, quenching of metabolic activity, extraction of relevant metabolites, and quantification.

Metabolomics research occurred in the late 1960s. So in the early 1970s, first reports appeared dealing with the comparison of various methods for extraction and their effect on observed concentrations. Eukaryotic yeast cells were examined [79] as well as *E. coli* cells and first conclusions about metabolic regulations were drawn [80]. To this end, single metabolite concentrations were determined by fluorometric–enzymatic analyses. Already in these early days, questions were dealt with that still are of interest today—like searching for the best extraction method or filtration for separation of cells and culture medium. The term 'metabolome' to describe the total metabolite pool has only been used since the late 1990s. At that time, semiquantitative methods such as two-dimensional thin-layer chromatography were applied to determine as many metabolites as possible with one method and to identify differences in metabolic patterns under varying growth conditions [81].

Today, metabolomics are an important tool to enable and verify targeted strain development by contributing to a detailed analysis and understanding of cellular metabolism and its regulation and thereby complementing genomics, transcriptomics, proteomics, and fluxomics [82]. Most approaches deal with the analysis of the central metabolism, but as precursors and energy equivalents from there are critical for all secondary metabolites—to which group most biotechnologically derived products belong—their analysis has the potential to give insight into these parts of metabolism as well [83]. Moreover, many fast regulatory mechanisms such as allosteric or feedback inhibition can virtually only be investigated in vivo via metabolite concentrations. Knowledge and understanding of these mechanisms is crucial to get a deep and comprehensive insight into metabolic processes [84]. Further, first steps have been taken into the direction of single-cell metabolomics [85]. So far, these approaches are limited to big cells such as neurons and the initial development is just in progress.

Several distinct work routines are applied nowadays for metabolite quantification. Figure 1 depicts various possibilities to combine the steps necessary for successful comprehensive metabolome analysis, which are discussed in detail below. These steps strongly influence each other and are under constant development and evaluation among the groups working in this field. Despite huge improvements, no ultimate method that fulfills all needs is available for any of these steps, let alone for the whole procedure.

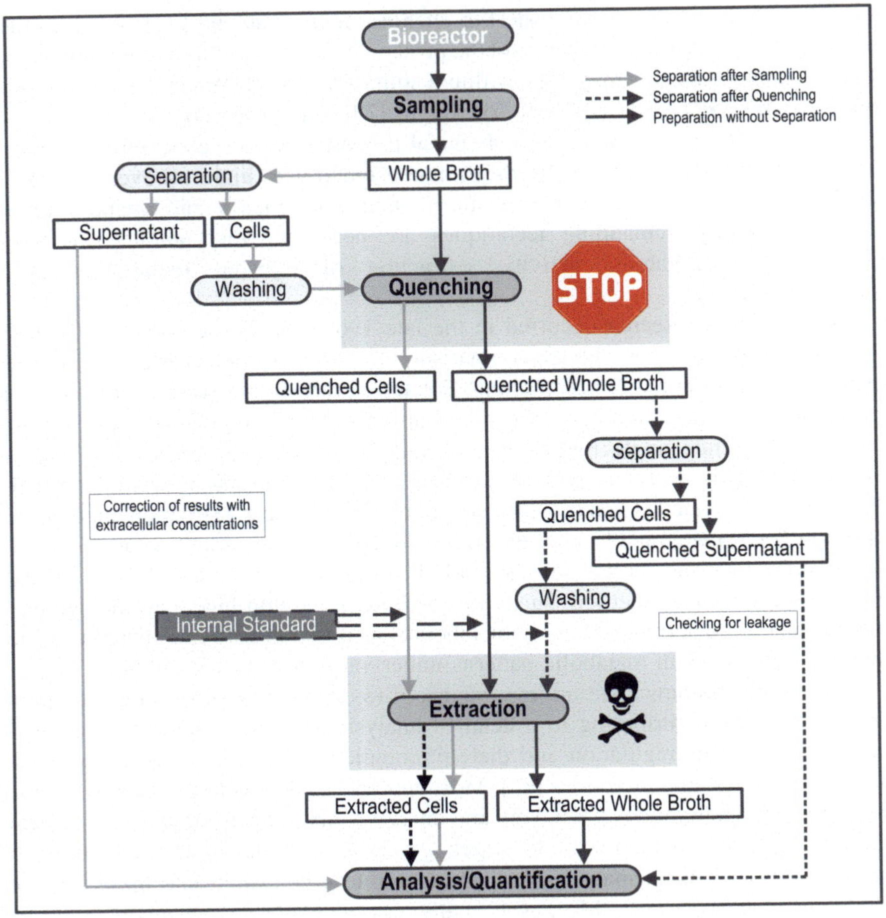

Fig. 1 Schematic summary of multiple sample preparation methods for quantitative analysis of intracellular metabolites

4.1 Sampling

The most important requirement about sampling in the context of metabolomics is that it has to be performed very fast to capture the metabolic state of the cells in the reaction system unbiased. The small metabolite pools inside the cells (nM–mM) are consumed rapidly by enzymes with high turnover rates (mM s^{-1}) so the time frame for a metabolic snapshot lies below 1 s [86]. This leads to the necessity to quench and/or extract the cells during or immediately after sampling and to store the samples appropriately afterwards.

A lot of sampling devices have been developed to cover various applications and to make different organisms available for metabolic analyses, each with its own advantages and disadvantages [87]. Many of them are directly linked to the reactor by a valve and can be used for sampling from steel reactors by channeling the sample into quenching fluid [88] or into a heat exchanger [89]. Only few devices are available for rapid sampling from glass reactors. The one developed by [90] requires manual handling but allows parallel sampling and inactivation by dispersing the sample in quenching solution. Different demands are met by a device that allows anaerobic rapid sampling from a culture flask [91]. To minimize manual handling, developments including automated timer and valves [92] or continuous sampling and inactivation [93] were adopted. An increasing number of fully automated devices enable high sampling frequencies (>5 s^{-1}) by providing a magazine of sampling tubes that moves under the sampling port [94]. This minimizes manual handling and allows for a detailed time resolution of the metabolic status.

A special solution for parallel perturbation and sampling is the BioScope reactor [16]. This small plug-flow reactor is connected to a laboratory fermenter. Broth from the reactor and perturbing agent are mixed and then flow along serpentines. Several sequential ports allow sampling at different times after perturbation.

A distinct method for sampling is fast filtration. It is clearly slower (several seconds) than the other sampling devices but leads to immediate separation of intra- and extracellular metabolites [95]. Moreover, in contrast to all other in-house built methods, no special equipment is required [96]. Nevertheless, the time required for sampling hinders the detection of metabolites with high turnover rates, e.g., from glycolysis. New approaches combine fast filtration and automated sampling to reduce the time required for sampling [97].

4.2 Quenching

Quenching is the step to instantly stop or 'freeze' the metabolism to avoid consumption of metabolites by ongoing metabolic activities. This step must take place simultaneously to or immediately after sampling.

A common method for quenching is the dispersion of cell suspension in a cold methanol solution [98], but also high temperatures and high or low pH have been applied, each leading to individual advantages and disadvantages [83]. A big problem concerning quenching is cell leakage, which describes the cell damage during quenching and the resulting loss of metabolites from the cells into the supernatant. This leads to quantification mistakes and is a topic that especially occurs when quenching of prokaryotic cells is performed [99]. Several distinct methods for quenching bacteria, including various combinations of cold methanol and buffers as well as liquid nitrogen, have been described to lead to significant leakage of metabolites [100]. Eukaryotic cells such as yeasts or fungi are less susceptible for leakage due to their different cell structure and more robust cell wall, but it has been described for these organisms as well [101]. Their robustness on the other hand makes subsequent extraction more laborious, and moreover, the compartments in the cells cause different kinds of problems when doing metabolomics [102].

If leakage does occur, the separation of cells and medium after quenching leads to huge mistakes during analysis. In this case, a whole-broth extraction and the subsequent correction by considering extracellular concentrations should be performed [95]. These concentrations are usually determined by the analysis of a 'normal' sample of the supernatant without quenching. Since the extracellular space largely exceeds the intracellular space, even small extracellular concentrations can be enough to mask intracellular concentrations [100]. This is especially critical as even metabolites that usually are supposed to occur exclusively inside cells were detected extracellularly during batch cultivations of various organisms [103].

An approach to avoid quenching is rapid filtration to separate cells and medium directly after sampling. Filtration is immediately followed by a washing step and extraction. This method is applicable up to moderate biomass concentrations and circumvents the problems that arise from quenching, allows discrimination between intra- and extracellular metabolites, and lowers the salt concentration in the samples [96]. But as time passes before extraction is performed, in which the cells are still metabolically active, the probable loss of metabolite pools or changes in metabolome is comparatively high [78]. However, by choosing an appropriate washing solution with physical–chemical properties similar to the culture medium, this effect can be minimized [95].

4.3 Extraction

The aim of the extraction is to release all metabolites from the cells without destroying or modifying any of them. Moreover, extraction must lead to sample compositions that can be analyzed, e.g., a low salt concentration is preferable for the analysis by mass spectrometry to avoid effects of ion suppression [104]. In general, it is either possible to extract only the cells (and therefore separate cells and medium prior to extraction) or cells and medium ('whole-broth' extraction).

The extraction step is often performed by boiling the quenched cells in (buffered) water or ethanol. Further possibilities are the application of extreme pH or freeze–thaw cycles [98]. Also, mechanical extraction by deep freezing and grinding or sonication has been described [105]. Several results suggest that the conduction of various extraction methods in parallel would enable the analysis of the widest selection of different metabolites. By doing so, it is possible to consider the specific stabilities of metabolites when facing heat, acid, alkali, or (non)polar solvents [83]. For example, acidic acetonitrile extraction is described to improve quantification of triphosphates [102]. However, performing several extraction methods complicates and amplifies the experimental setup as well as the analytical extent.

Prior to analysis, concentration of the samples by lyophilization and resuspension in a small volume can compensate the diluting steps in quenching and extraction [106]. This additional step again increases the possibility to loose or modify metabolites during sample preparation, but with analytic techniques and devices becoming more and more sensitive, it can often be omitted.

4.4 Quantification

In order to get as much information as possible, the quantification of intra- and extracellular metabolites is necessary:

The quantification of extracellular metabolites can be realized relatively easily by standardized methods such as HPLC or enzymatic assays. Sample preparation is limited to centrifugation at low temperatures and/or filtration. Nevertheless, assuring a fast cool down of the extracellular samples is necessary to guarantee unbiased measurements of the supernatant. This can be achieved with pre-cooled sampling tubes filled with glass beads to facilitate heat transfer.

On the other hand, the exact quantification of intracellular metabolites poses challenges that must be faced with special techniques because a normal external calibration cannot be applied. There are two main reasons for that. First, sample preparation for the analysis of the metabolome includes extreme conditions, especially during quenching and extraction. It is not possible to state which metabolites are changed or destroyed to what extent during these steps, but it is sensible to assume that metabolite concentrations are reduced during preparation. The second point concerns matrix effects that occur during analysis. When applying mass spectrometry, ion suppression is a common phenomenon [104]. This means that nonvolatile components, i.e., salts, impede the ionization of bigger molecules and thus lead to false conclusions concerning their concentrations. Ion suppression is a problem during metabolome analysis because culture media contain high salt concentrations and this salt often cannot be eliminated during sample preparation.

The first issue can partially be faced by adding stabilizing reagents to the quenching solution, which improves the quantification of a critical group of metabolites [107]. The addition of an internal standard after sampling permits exact quantification of metabolites despite degradation during sample preparation and matrix effects. The internal standard's concentration has to be known, and it needs the same chemical and physical properties as the metabolite of interest, which leads to identical behavior during sample preparation. Nevertheless, it must be distinguishable from the metabolite during analytics. All these requirements are met by uniformly ^{13}C-labeled cell extracts that allow the application of isotope dilution mass spectrometry (IDMS) [108, 109]. The isotope solely influences the metabolite masses, so all other physical and chemical properties are maintained, but a separate quantification of the standard via mass spectrometry is possible. The quantification of the metabolites is then realized by comparison of signal ratios.

A different approach for exact quantification via an internal standard is standard addition. The sample is split and spiked with different known concentrations of the component of interest. A comparison of signal intensities from spiked samples and a sample without standard addition allows the identification of the component's concentration. This method does not require U-^{13}C labeling but multiplies the number of samples. It can be used for characterization of a labeled U-^{13}C cell extract, which is then applied for IDMS [96].

4.5 Analytics

The ideal technique for metabolome analysis would be able to identify and quantify all metabolites simultaneously in a small sample amount, within a short time and without interference with the sample matrix.

In the early days of metabolomics, enzymatic assays were used to quantify metabolites, one at a time [79]. This technique was limited by the necessary sample volumes and the detection limits. Further, an early method for identification of metabolites was two-dimensional thin-layer chromatography, which also has limited capacities and is dependent on laborious labeling [110]. These historical techniques have been replaced by modern approaches that enable analysis, exact quantification, and identification of many metabolites at the same time and thus are closer to the ideal technique.

Today, quantitative analyses are mostly performed by mass spectrometry that works on the basis of mass to charge ratio (m/z). The benefits of mass spectrometry are a high sensitivity, the simultaneous quantification of many different metabolites from various classes and the small sample amount required [111]. Mass spectrometry comprises the steps ionization, ion separation, and ion detection. Various techniques are available for each of these steps, and the suitable ones for the quantification of metabolites must be selected. Next to a soft ionization technique such as electrospray ionization (ESI) or matrix-assisted laser desorption ionization (MALDI) that prevents damage of the metabolites, the ion separation method must be capable of handling rather small molecules. A triple quadrupole ensures this and moreover enables an especially sensitive quantification because only selected ions are measured at a certain time [111]. Another widespread technique is time of flight (TOF) [112].

This multitude of analytical approaches is even broadened by the possibility to link the mass spectrometer to a liquid chromatography (LC), gas chromatography (GC), or capillary electrophoresis (CE), depending on the properties of the metabolites of interest. These facilitate analysis by separating the metabolites according to their chemical or physical properties prior to mass spectrometry. LC is the most common choice because it is robust and suitable for quantification for the largest set of metabolites without requiring pre-analysis sample derivatization [113]. The application of UHPLC or μLC lowers the sample amount required while leading to shorter measurement times and better separation [111]. The combination of UHPLC-MS/MS is, for example, applied in a method that allows the quantification of 138 metabolites from various classes within 30 min, including the separation of isomers [114]. But although the development of analytic devices and techniques now allows much more precise quantification than some years ago, it is still a distant goal to measure, e.g., all 1136 metabolites that are included in the latest *E. coli* model [53] with one single method.

A different analytical approach is untargeted profiling that aims at identifying as many metabolites as possible rather than quantifying them exactly. This is also called 'footprinting' and 'fingerprinting' when referring to the exo- and the endometabolome, respectively [83]. These approaches are mainly realized by applying nuclear magnetic resonance (NMR) methods because they are especially capable of qualitative analyses and structure characterization. Further advantages of NMR are nondestructiveness and short measurements times (seconds–minutes) that enable high-throughput applications. One drawback is the reduced sensitivity in comparison with MS [115]. Nevertheless, NMR—next to fluorescent assays that are limited to single metabolites—is potentially capable of measuring metabolites inside cells without the need to quench and extract. But of course sensitivity and especially the detection of rare metabolites are major challenges [78].

Another possibility to minimize sample preparation is direct infusion of samples from microtiter plates [116] or agar plates [117] into a mass spectrometer. These techniques are adequate for high-throughput screening but not for exact quantification of metabolites.

5 Perturbation Experiments During Microbial L-phenylalanine Production

In this chapter, a specific example is given to demonstrate experimental applicability of perturbation experiments. Within this example, steady-state perturbation experiments are used for the characterization of cell metabolism during a microbial L-phenylalanine production process.

5.1 Industrial Relevance of Microbial L-phenylalanine Production

L-phenylalanine (L-phe) is an essential aromatic amino acid. It is synthesized via the aromatic amino acid pathway that also leads to L-tyrosine (L-tyr) and L-tryptophan (L-trp). The first step of the common pathway is an aldol reaction of erythrose-4-phosphate (E4P) from the pentose phosphate pathway and phosphoenolpyruvate (PEP) from glycolysis catalyzed by DAHP synthase. In the following steps, one more PEP is consumed for every synthesized L-phe, L-tyr, and L-trp. The pathway is tightly regulated by feedback inhibition of DAHP synthase isoenzymes by the final products [118].

L-phenylalanine is used widely in food and feed industry as an additive, but its high industrial relevance is caused mainly by its usage as a precursor for the artificial sweetener aspartame, which is around 170 times sweeter than sugar [119]. As some people who have phenylketonuria should be cautious with phenylalanine containing substances, soft drink bottles with 'light' or 'diet' soda carry a warn sign 'contains a source of L-phenylalanine.'

The microbial production of L-phenylalanine in fed-batch or continuous processes has been studied for several decades with special focus on *E. coli* [120–122] and *Corynebacterium glutamicum* [123]. Important targets for genetic modifications to increase the production of aromatic amino acids are the improvement of precursor supply (E4P and PEP) and the reduction of feedback inhibition. Precursor supply is mostly influenced by overexpression or deletion of relevant enzymes [124], while inhibition is tackled with feedback-resistant enzymes [125]. Another widespread feature of L-phenylalanine-producing strains is L-phe and L-tyr auxotrophy [119]. The most common carbon source in the studies about L-phenylalanine production is glucose. High product yields and concentrations are achieved, for example, 18.6 mol % and 34 g L^{-1} L-phe in a controlled fed-batch process [126]. Applying in situ product recovery enhanced product concentration to 46 g L^{-1} [126]. Also, other sources state high product concentrations of up to 50 g L^{-1} [125, 127–129].

Glycerol is an attractive alternative carbon source for L-phenylalanine production. Next to its low price and high availability as a by-product of biodiesel production [130], its reduction potential and theoretical L-phenylalanine yield are higher in comparison with glucose. Moreover, in contrast to glucose, glycerol is transported into the cells without consumption of PEP; this feature is expected to be beneficial for precursor supply [131]. Various *E. coli* strains have been developed for optimized L-phenylalanine production from glycerol. Novel approaches including transporters for substrate and product [132] complement the established genetic modifications for precursor supply and minimal inhibition [131]. Recently, we described a fed-batch process with glycerol as sole carbon source throughout the whole process with a product concentration of 21.3 g L^{-1} [133]. Since the achieved L-phenylalanine concentration is below other results, the potential for improvement is obvious. The methods applied are presented here to gain insight into L-phe metabolism of *E. coli* cells in the fed-batch production process and thereby find targets for optimization.

5.2 L-phenylalanine Production Process

Microbial L-phenylalanine production on glycerol was intensively studied in the last years. Therefore, different recombinant strains were engineered and characterized recently [131, 133, 134]. Detailed description about strain design and modification of the recombinant strain presented here, *Escherichia coli* FUS4 $pF81_{kan}$, are available in several publications [119, 126, 131, 133, 134, 135]. Beside strain improvement, further work has been done to improve process design for microbial L-phenylalanine production. Therefore, a fed-batch process strategy has been presented with glycerol as carbon source during L-phenylanine production [134]. Process design significantly influences applicability of metabolic perturbation experiments. Because of this reason, some central aspects have to be considered for process control and process strategy:

- Utilization of defined mineral salts media to ensure controlled substrate uptake
- Substrate-limited process design (stirred-tank bioreactor in fed-batch or continuous mode) to control substrate uptake and the metabolic state of the cells
- Balanced and fully quantifiable extracellular carbon fluxes (e.g., substrate uptake, production of side products and main product, respiration component CO_2)

To meet these criteria, a controlled substrate-limited L-phenylalanine production process was established on a 15 L-scale with a defined salt medium [133]. Exemplary results of one cultivation are shown in Fig. 2. The fed-batch process was divided into two process phases (biomass formation and production). Within the first process phase, biomass formation was realized by a growth-rate-controlled [136] exponential glycerol supply ($\mu_{set} = 0.1\ h^{-1}$). By doing so, a biomass concentration of 28.6 g L^{-1} was achieved after 43.4 h (see Fig. 2a). Furthermore, L-phenylalanine and L-tyrosine were added during this phase for supplementation of strain-specific auxotrophies [133]. The first reaction of the amino acid pathway, catalyzed by DAHP synthase, is inhibited already at low concentration of L-tyrosine [137]. Because of this enzymatic property, lowest possible residual concentration of L-tyrosine is favoured. For this purpose, L-tyrosine concentration in feeding media was adapted iteratively within multiple cultivations. This optimized media composition resulted in low L-tyrosine concentration (0.27 g L^{-1}) at the end of the growth phase (see Fig. 2b). Furthermore, residual L-tyrosine concentration was completely metabolised during the first hours of the second process phase (production phase). As a consequence, inhibition of L-phenylalanine synthesis by L-tyrosine was successfully excluded.

After addition of IPTG (0.3 mM), production of L-phenylalanine was initiated at the beginning of the second process phase. During this phase, amino acid formation was realized by a constant supply of glycerol (0.18 $g_{glycerol}\ g_{CDW}\ h^{-1}$). Ammonia supply was ensured by pH control (titration agent: 25 % NH_4OH) and ammonia

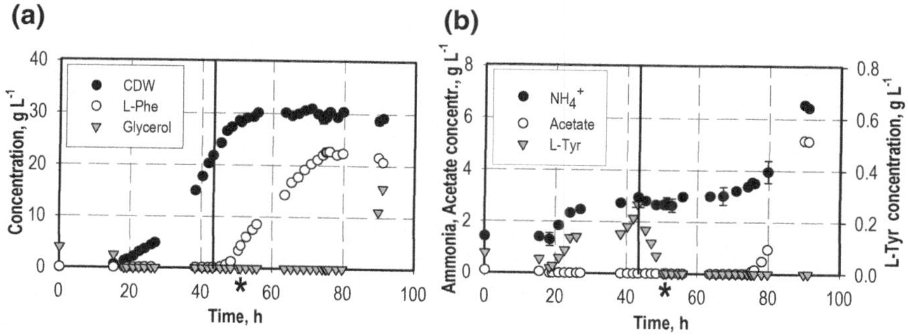

Fig. 2 L-phenylalanine fed-batch production with *E. coli* FUS4 pF81$_{kan}$. *Vertical lines* (43.4 h) indicate IPTG addition for the induction of protein expression. Illustration: Concentrations of cell dry weight (CDW), L-phenylalanine (L-phe), L-tyrosine (L-tyr) glycerol, ammonia (NH_4^+), and acetate. *Black asterisk* at 50.5 h indicates process time of biomass transfer for perturbation experiments

salts in feeding media. Constant ammonia concentration during production phase (see Fig. 2b) indicates an optimized ammonia supply for this fed-batch process. After about 30.6 h of production, a maximal L-phenylalanine concentration of 22.8 g L^{-1} was measured (see Fig. 2b). The highest biomass-specific production rate was 34 $mg_{L\text{-phe}}\ g_{CDW}^{-1}\ h^{-1}$, detected at about 50 h. At 75 h, initial by-product formation of acetic acid set in, which increased until the end of the cultivation (see Fig. 2b).

For applicability of perturbation experiments, the necessity of complete extracellular fluxome quantification is given. Therefore, carbon balances are a proper tool to evaluate whether all relevant extracellular fluxes are considered during the production process. In Fig. 3, a carbon balance of the presented L-phenylalanine production process with *E. coli* FUS4 $pF81_{kan}$ is shown. Carbon balance is closed (>94 %) throughout the whole process. This balance indicates that the applied analytical methods are sufficient for extracellular flux estimation and that a substrate-limited process control was established successfully. As a consequence, criteria for applicability of perturbation experiments were fulfilled for this 15 L-scale production process.

5.3 Parallelized Batch Steady-State Perturbation Experiments During L-phenylalanine Production

As mentioned in Sect. 2.2, a constant enzyme level is necessary for direct comparability of fluxome and metabolome data from multiple steady-state perturbation experiments. This criterion is realized mostly by sequentially applying several

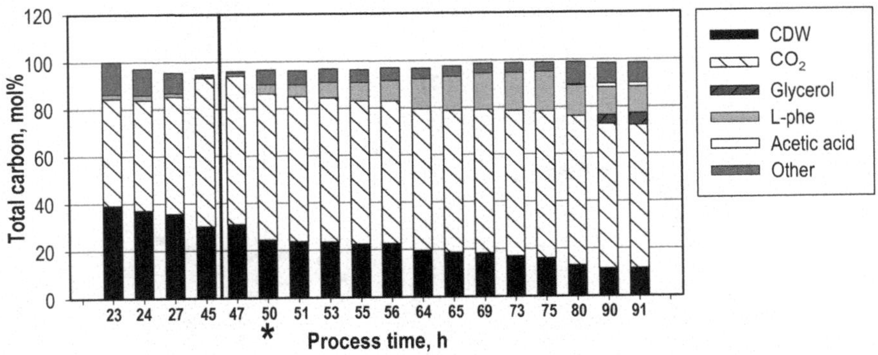

Fig. 3 Carbon balances for the fed-batch L-phenylalanine production process with *E. coli* FUS4 $pF81_{kan}$. Balances are standardized to 100 % at the end of the initial batch phase. Substances considered: Biomass (CDW), CO_2, glycerol, L-phenylalanine (L-phe), and other (lactic acid, citric acid, L-glutamate, L-tyrosine, carbon eliminated by sampling or for perturbation experiments). *Vertical line* indicates IPTG addition for the induction of protein expression. *Black asterisk* indicates process time of biomass transfer for perturbation experiments

perturbations to cells from stationary chemostat cultivations or growth-rate-controlled fed-batch processes. Unfortunately, neither chemostat mode nor growth-rate control is practically relevant in industrial production processes.

Since constant enzyme levels also cannot be ensured during the production phase of the herein presented L-phenylalanine production process, sequential application of steady-state perturbation experiments is not realizable. Because of this, the method of 'Rapid Media Transition,' proposed by [19] for steady-state perturbation experiments in parallel to a substrate-limited fed-batch process, was extended and will be described below.

To enable comparability of perturbation data, we modified the experimental setup of 'Rapid Media Transition.' Therefore, emphasis was put on comparability of enzyme levels within all perturbation experiments. In contrast to single-perturbation experiments as proposed by [19], we performed multiple perturbation experiments at the same time in parallel to the production process. This was realized by using a commercially available fourfold parallel stirred-tank bioreactor system (DASGIP technology/Eppendorf, Germany), operating on a 0.5 L-scale in parallel to the 15 L-scale L-phenylalanine production process. The bioreactor system was equipped with online probes for pH and dissolved oxygen monitoring as well as peristaltic pumps for two-sided pH control (pH 7 using 5 M NaOH or 28 % H_3PO_4). Overhead-driven Rushton impellers and gas-mixing units (25 % O_2, 4 L min^{-1}) were used for oxygen supply. Oxygen consumption and CO_2 production were monitored online in the off-gas (Easy Line, ABB Automation, Zurich Switzerland). Further, all four analytical reactors were equipped with a modified, improved version of the rapid sampling device described by [90].

A schematic overview of parallelized batch steady-state perturbation experiments is shown in Fig. 4. For perturbation experiments, 3.6 L culture broth was withdrawn from the L-phenylalanine production process. Cells were separated by centrifugation (3260 g, 37 °C, 7.5 min). The supernatant was discarded, and biomass was resuspended in 400 mL minimal media (37 °C). After determination of the total volume of resuspended cells, volume was divided equally for simultaneous inoculation of the four perturbation reactors (each prepared with 400 mL of preheated minimal media). The perturbation was induced by four different carbon sources, which were present in one of the perturbation reactors, respectively. In the here presented batch perturbation experiments, pyruvate, glucose, succinate, and glycerol were used as perturbation substrates. Perturbation was performed for 19 min after biomass transfer. Samples of extracellular metabolites and biomass concentration were taken simultaneously from all stirred-tank bioreactors at four times during the short-term cultivation (1, 7, 13, and 19 min). For one exemplary set of perturbation experiments, substrate concentration for all four perturbation reactors is shown in Fig. 5. Cell harvest from the 15 L-scale L-phenylalanine production process was performed at the time of highest production rate (process data not shown; process performance comparable to Fig. 2). All carbon sources were immediately metabolized by the cells. Linear decrease of substrate concentration was detected in all analyses, indicating metabolic steady state. The maximal biomass-specific uptake rate of the different substrates was estimated by linear regression and is listed in Table 1. Linear

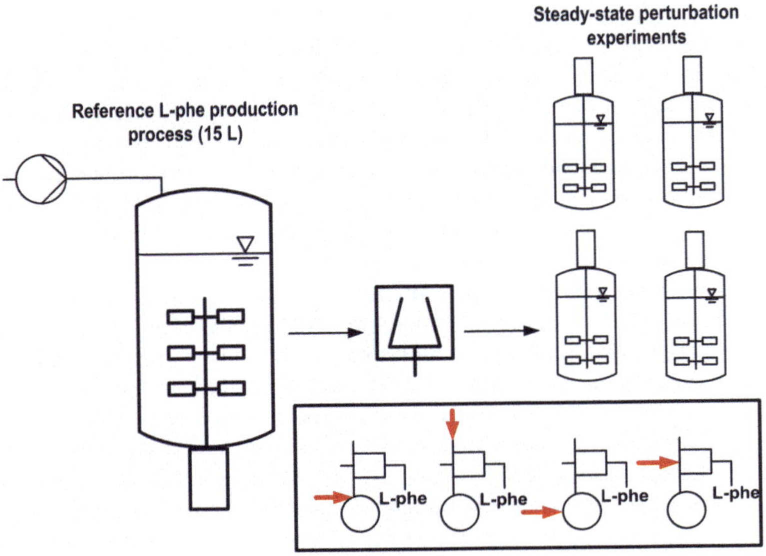

Fig. 4 Applied setup of 'Rapid Media Transition' for parallelized batch perturbation experiments. Biomass from the production process is transferred to four laboratory-scale stirred-tank reactors, which operate in batch mode with different carbon sources. Cells are separated by a centrifuge

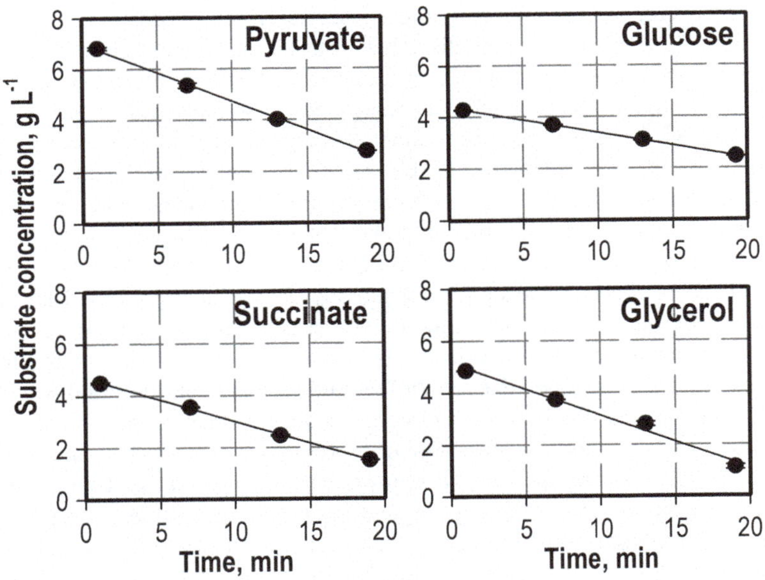

Fig. 5 Substrate concentrations during batch perturbation experiments with four different substrates (pyruvate, glucose, succinate, and glycerol). *Black dots* measured concentrations; *black line* regression analysis

Table 1 Extracellular fluxes (unit: mmol g_{CDW}^{-1} h^{-1}) of substrate uptake, production rate, by-product synthesis and respiration during four parallel batch steady-state perturbation experiments on pyruvate, glucose, succinate, and glycerol

	Pyruvate	Glucose	Succinate	Glycerol
Substr.	−5.14 ± 0.07	−1.02 ± 0.01	−3.04 ± 0.03	−4.57 ± 0.02
OUR	−4.86 ± 0.19	−3.03 ± 0.20	−9.79 ± 0.17	−4.86 ± 0.10
CER	6.91 ± 0.17	3.08 ± 0.13	11.79 ± 0.55	3.47 ± 0.10
L-phe	0.06 ± 0.00	0.14 ± 0.01	0.06 ± 0.04	<0.01
By-products				
Acetate	0.98 ± 0.01	n.d.	n.d.	1.20 ± 0.01
Malate	n.d.	n.d.	0.46 ± 0.01	n.d.
Lactate	n.d.	n.d.	n.d.	1.01 ± 0.01
Pyruvate	–	n.d.	0.48 ± 0.01	n.d.

OUR biomass-specific oxygen uptake rate; *CER* biomass-specific CO_2 production rate

by-product formation was detected in three experiments (on pyruvate: acetate; on succinate: malate and pyruvate; on glycerol: lactate and acetate; detailed in Table 1). Since sufficient oxygen supply was ensured by enriched air-mixing and dissolved oxygen concentration was monitored online (>5 %), oxygen limitation can be excluded, also there was a lack of formate, a typical product of *E. coli* anaerobiosis. Because of this reason, the measured by-product formation can be attributed to an effect of overflow metabolism during the performed batch cultivations. L-phenylalanine production was detected in all experiments (see Table 1). Extracellular fluxes of respiration were considered by the calculation of biomass-specific O_2 uptake rate (OUR) and biomass-specific CO_2 production rate (CER) from online measured off-gas data. OUR and CER data for the parallel performed perturbation experiments are shown in Fig. 6. Within about six minutes, OUR and CER indicate a constant value for all experiments, which is stable until the end of the cultivation.

As described by [19], metabolic steady state is indicated by these stationary extracellular fluxes of substrate uptake and respiration. This means that the method of steady-state perturbation experiments was applied successfully for cells from an L-phenylalanine production process. Comparability of perturbation data was ensured by a parallelized setup of multiple single-perturbation experiments. Additionally, we were able to demonstrate in accordance with [19] that *E. coli* is able to metabolise the applied carbon sources immediately and that extracellular fluxes are constant within a few minutes after the cells are perturbed. Since a new metabolic steady state was detectable within less than six minutes, application of more than one perturbation per reactor appeared practicable within the available analytical time of 19 min. As a consequence, the method of parallelized steady-state perturbation experiments was further modified and will be described in the following chapter.

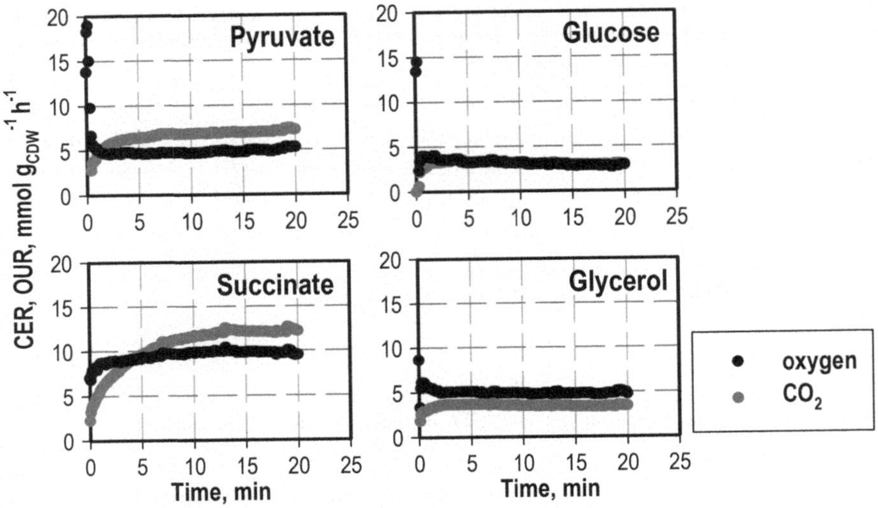

Fig. 6 Biomass-specific oxygen uptake rate (OUR; *black lines*) and the biomass-specific CO_2 production rate (CER; *gray lines*) during perturbation experiments with different substrates (pyruvate, glucose, succinate, glycerol)

5.4 Parallelized Fed-Batch Steady-State Perturbation Experiments During L-phenylalanine Production

Since metabolic steady state was achieved within less than six minutes in the applied parallelized batch perturbation experiments, further modifications of the experimental setup were reasonable. Perturbation experiments using the method of 'Rapid Media Transition' are bounded by the applicable cultivation time. Since a constant enzyme level can only be guaranteed for about 30 min after cell harvest from production process [106], perturbation experiments are limited to a timescale of about 20 min after biomass transfer into analytical scale. Within this given analytical timescale, three different metabolic steady states (3 × 6 min) are likely to be achieved. For sequential adjustment of different defined perturbations within one analytical reactor, substrate-limited conditions have to be realized. Targeted perturbation for each new metabolic steady state can only be realized by avoiding substrate excess conditions in the previously applied metabolic steady state. To realize this essential criterion, parallelized perturbation experiments were performed in fed-batch mode. By doing so, also effects of overflow metabolism, which are presented in batch perturbation experiments, may be excluded.

The new schematic overview is given in Fig. 7. By the usage of a fed-batch profile with three stepwise increasing (7, 13 min) supply rates (A, B, C), three different substrate uptake rates were aimed (see Fig. 8). To avoid substrate excess, feeding profile was adjusted for each carbon source, according to the measured

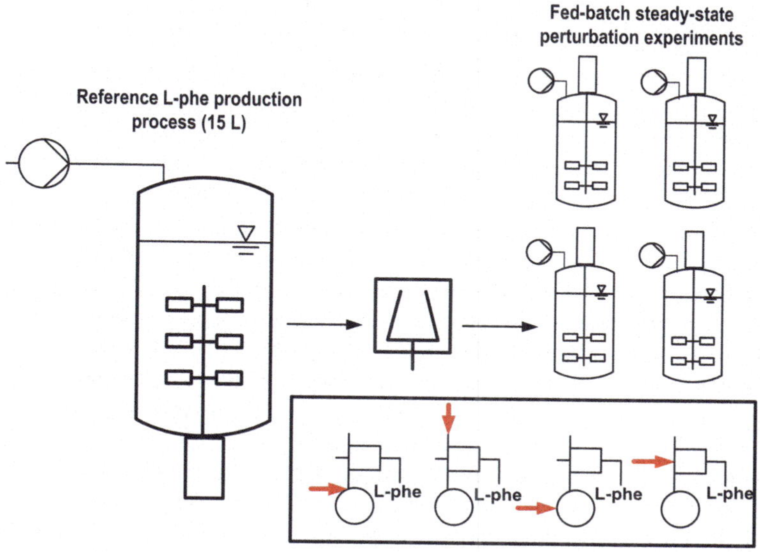

Fig. 7 Applied setup of parallelized fed-batch perturbation experiments. Biomass from the production process is transferred to four laboratory-scale stirred-tank reactors, which operate in substrate-limited fed-batch mode with different carbon sources. Cells are separated by a centrifuge

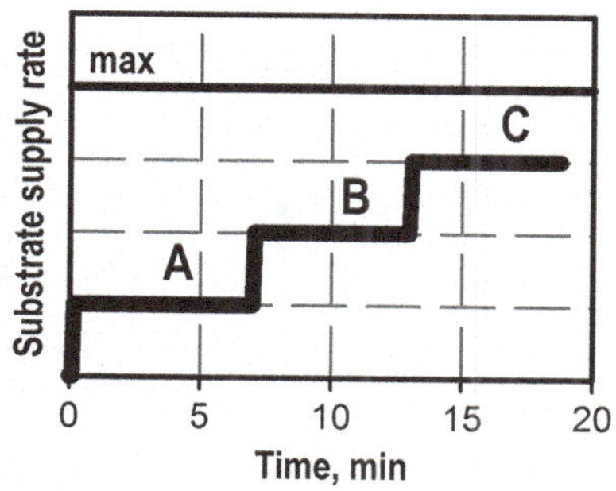

Fig. 8 Applied substrate feed profile during parallelized fed-batch perturbation experiments. *Black horizontal line* maximal biomass-specific uptake rate

maximal specific substrate uptake rate, achieved during batch experiments (see Table 1). Feeding was realized by peristaltic pumps, equipped with scripted software control. Further, feed-media supply was monitored online using four analytical balances. Substrate concentration in feed media was quantified off-line using HPLC. Substrate limitation during perturbation experiments was controlled by taking samples of extracellular metabolites simultaneously from all stirred-tank bioreactors at four times during the short-term cultivation (1, 7, 13, and 19 min). Furthermore, L-phenylalanine and biomass concentration were measured off-line.

By doing so, extracellular fluxes of substrate uptake, production rate, and respiration were calculated.

For one set of perturbation experiments, extracellular fluxes of substrate uptake, production rate, and respiration for the three steady-state conditions (A, B, C) and for all four perturbation reactors are shown in Fig. 9 and listed in Table 2. Cell harvest from the 15 L-scale L-phenylalanine production process was performed at the time of highest production rate (50.5 h see Fig. 2). Substrate limitation was successfully achieved during all steady-state perturbation experiments, and no by-products were detected. Three different distributions of extracellular fluxes were realized by applying these substrate-limiting conditions. For all carbon sources, three different substrate uptake rates were adjusted successfully (see Fig. 9). As predefined by using the rationally designed feed profile, highest substrate uptake rate was below maximal specific uptake rate, determined in batch experiments (compare Tables 1 and 2). Perturbation of L-phenylalanine production rate was also successful. Whereas high L-phenylalanine production rates of up to 0.31 mmol g_{CDW}^{-1} h^{-1} were achieved for highest glycerol uptake, only small fluxes (<0.02 mmol g_{CDW}^{-1} h^{-1}) were measured during lowest substrate supply of all carbon sources. In Fig. 9, the three extracellular flux distributions on glycerol during metabolic perturbation experiments are compared to the present flux distribution in 15 L-scale at the time of perturbation experiments. All extracellular flux values of the 15 L-scale process are between the highest perturbation steady state (C) and the

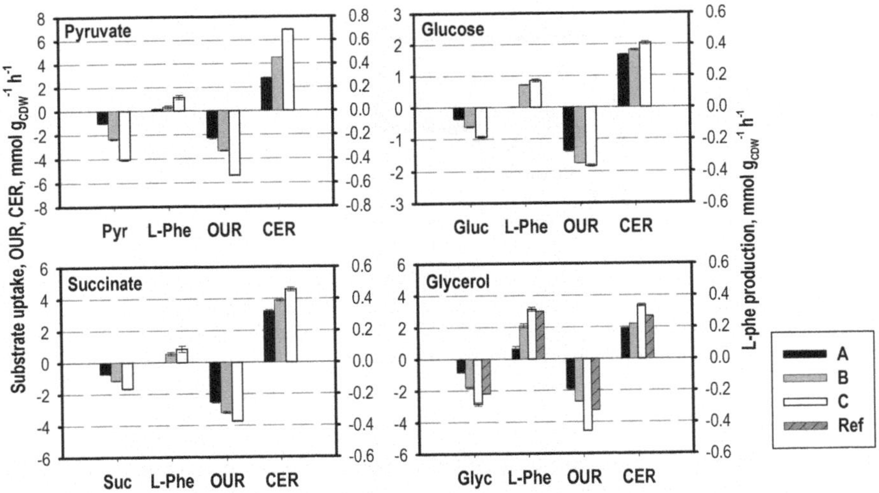

Fig. 9 Extracellular fluxes (unit: mmol g_{CDW}^{-1} h^{-1}) of substrate uptake, production rate, and respiration during four parallelized fed-batch steady-state perturbation experiments on pyruvate, glucose, succinate, and glycerol. *OUR* biomass-specific oxygen uptake rate; *CER* biomass-specific CO_2 production rate. *A*, *B*, *C* the three in sequence applied steady-state conditions for each carbon source. *Ref* extracellular fluxes of the 15 L-scale production process which was operating in parallel to the performed perturbation experiments

Table 2 Extracellular fluxes (unit: mmol g_{CDW}^{-1} h^{-1}) of substrate uptake, production rate, and respiration during four parallelized fed-batch steady-state perturbation experiments on pyruvate, glucose, succinate, and glycerol

	Pyruvate			Glucose		
	A	B	C	A	B	C
Substr.	−1.03 ± 0.02	−2.35 ± 0.05	−4.07 ± 0.08	−0.36 ± 0.01	−0.61 ± 0.02	−0.94 ± 0.03
OUR	−2.27 ± 0.10	−3.33 ± 0.04	−5.37 ± 0.04	−1.37 ± 0.03	−1.75 ± 0.01	−1.83 ± 0.03
CER	2.78 ± 0.03	4.50 ± 0.01	6.90 ± 0.04	1.67 ± 0.01	1.82 ± 0.02	2.03 ± 0.04
L-phe	0.01 ± 0.01	0.02 ± 0.01	0.11 ± 0.02	n.d.	0.14 ± 0.01	0.17 ± 0.01
	Succinate			Glycerol		
	A	B	C	A	B	C
Substr.	−0.73 ± 0.01	−1.17 ± 0.02	−1.66 ± 0.03	−0.82 ± 0.03	−1.81 ± 0.07	−2.86 ± 0.11
OUR	−2.50 ± 0.06	−3.19 ± 0.06	−3.72 ± 0.03	−1.89 ± 0.05	−2.69 ± 0.01	−4.55 ± 0.03
CER	3.22 ± 0.06	3.89 ± 0.06	4.58 ± 0.12	1.91 ± 0.02	2.19 ± 0.01	3.36 ± 0.07
L-phe	n.d.	0.05 ± 0.01	0.08 ± 0.02	0.07 ± 0.02	0.21 ± 0.01	0.31 ± 0.01

OUR biomass-specific oxygen uptake rate; *CER* biomass-specific CO_2 production rate. *A*, *B*, *C* the three in sequence applied steady states for each carbon source

intermediate steady state (B). This result clearly demonstrates that cell metabolism of the L-phenylalanine production was perturbed successfully in both directions. This bidirectional feature is essentially relevant for measured data-driven theoretical tools such as metabolic control analysis (MCA), which centrally accounts for perturbations around a certain metabolic state.

Formation of three metabolic steady states during these parallelized fed-batch perturbation experiments was again verified by off-gas measurements. The biomass-specific rate of oxygen uptake (OUR) for all cultivations is shown in Fig. 10. In particular, for experiments with glycerol and pyruvate as carbon source, a stepwise increasing off-gas profile is obvious, which correlates well with the applied feeding profile. Due to the reduced absolute values of oxygen uptake in experiments with glucose and succinate, off-gas profiles were not as definite as for pyruvate and glycerol. This point mainly results from the targeted smaller absolute changes in feeding profiles, which were applied for these two carbon sources. By consequence, the three metabolic states were also successfully realized within these experiments.

5.4.1 Intracellular Flux Estimation

Twelve different metabolic steady states were successfully realized by applying the proposed setup of parallelized fed-batch steady-state perturbation experiments

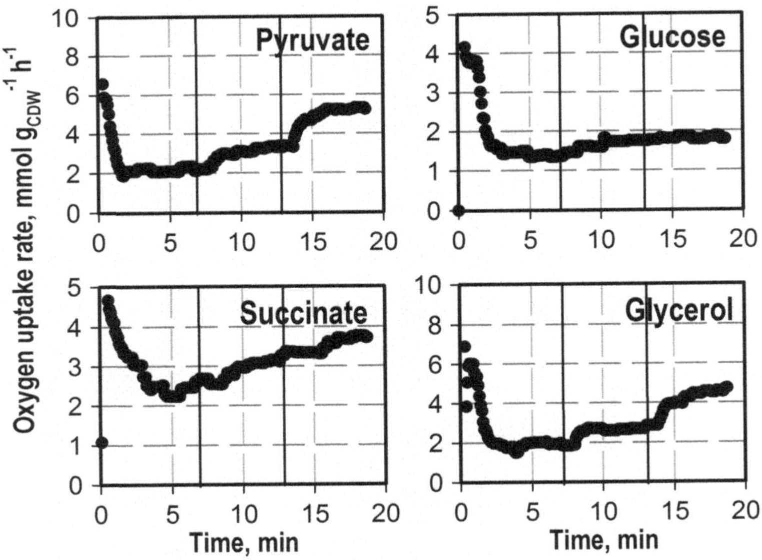

Fig. 10 Biomass-specific oxygen uptake rate during fed-batch perturbation experiments with different substrates (pyruvate, glucose, succinate, glycerol). *Vertical black lines* (7, 13 min) indicate end of steady state and time of increasing perturbation step according to feed profile (Fig. 8)

during an L-phenylalanine production process. As a result of this, 12 different extracellular flux distributions were measured. For the estimation of intracellular fluxes, constraint-based approaches (*g*FBA and FVA) and COBRA Toolbox v2.0 [62] in MATLAB 2014a (solver: TOMLAB cplex) were used. The estimations were based on the genome-scale *E. coli* model *i*AF1260 [52]. Further, the genome-scale model was adapted to consider present genetic modifications. The reaction of prephenate dehydrogenase was defined as a zero flux to consider tyrosine auxotrophy in *E. coli* FUS4 $pF81_{kan}$. The inactivation of additional enzymes improved the representation of cultivation conditions: tryptophanase and L-tryptophan uptake (no L-tryptophan in media), malic enzymes (inactive during L-phenylalanine production on glycerol [44]), glycerol dehydrogenase (inactive under aerobic conditions [138]), and fructose-6-phosphate aldolase (inhibited by glycerol, if present at all [139]). L-tyrosine uptake was allowed to enable biomass formation. For all flux estimations, substrate uptake and L-phenylalanine production rates were set as constraints and biomass formation as objective function. Exemplary flux distributions are discussed below and depicted in Figs. 11a, b and 12a, b. They originate from estimations with extracellular rates of the highest feeding level (C) during the applied parallelized perturbation experiments.

Pyruvate was taken up with 4.07 mmol g_{CDW}^{-1} h^{-1} (Fig. 11a). The majority of carbon was converted to AcCoA and then metabolized in the citric acid cycle and the glyoxylate bypass. PEP was generated directly from PYR (0.49 mmol g_{CDW}^{-1} h^{-1}) and from oxaloacetate (0.31 mmol g_{CDW}^{-1} h^{-1}). PEP was used for L-phenylalanine synthesis and for gluconeogenesis, which took place up to F6P. F6P and GAP were the origins of pentose phosphate pathway which led to the L-phenylalanine precursor E4P in the amount required for the constrained L-phenylalanine production rate (0.11 mmol g_{CDW}^{-1} h^{-1}).

Glucose as carbon source (Fig. 11b) was taken up via the PTS (0.94 mmol g_{CDW}^{-1} h^{-1}). The resulting G6P was metabolized primarily in glycolytic reactions, leading to 1.53 mmol g_{CDW}^{-1} h^{-1} PEP. Fractions of the available GAP and F6P were starting points for pentose phosphate pathway which generated E4P as a precursor of the measured L-phenylalanine production (0.17 mmol g_{CDW}^{-1} h^{-1}). The majority of PEP was necessary for the phosphorylation during glucose uptake, leading to PYR. Minor fluxes were active in the direction to OAA, directly to PYR and toward the biosynthesis of aromatic amino acids. The flux through citric acid cycle was around 0.90 mmol g_{CDW}^{-1} h^{-1}.

The uptake of succinate (1.66 mmol g_{CDW}^{-1} h^{-1}) led to highly active enzymatic reactions in the citric acid cycle and an inactive glyoxylate bypass (Fig. 12a). PEP was generated by the conversion of OAA (1.56 mmol g_{CDW}^{-1} h^{-1}) and was used primarily for the transformation to PYR and afterward AcCoA to keep the citric acid cycle running (1.04 mmol g_{CDW}^{-1} h^{-1}). Gluconeogenic reactions enabled the formation of GAP and F6P and consequently E4P in the pentose phosphate pathway. One molecule of E4P and two molecules of PEP were consumed for every molecule of L-phenylalanine (0.08 mmol g_{CDW}^{-1} h^{-1}).

The carbon source glycerol was taken up with 2.86 mmol g_{CDW}^{-1} h^{-1} during the highest feeding level in the metabolic analysis (Fig. 12b). Glycerol was converted

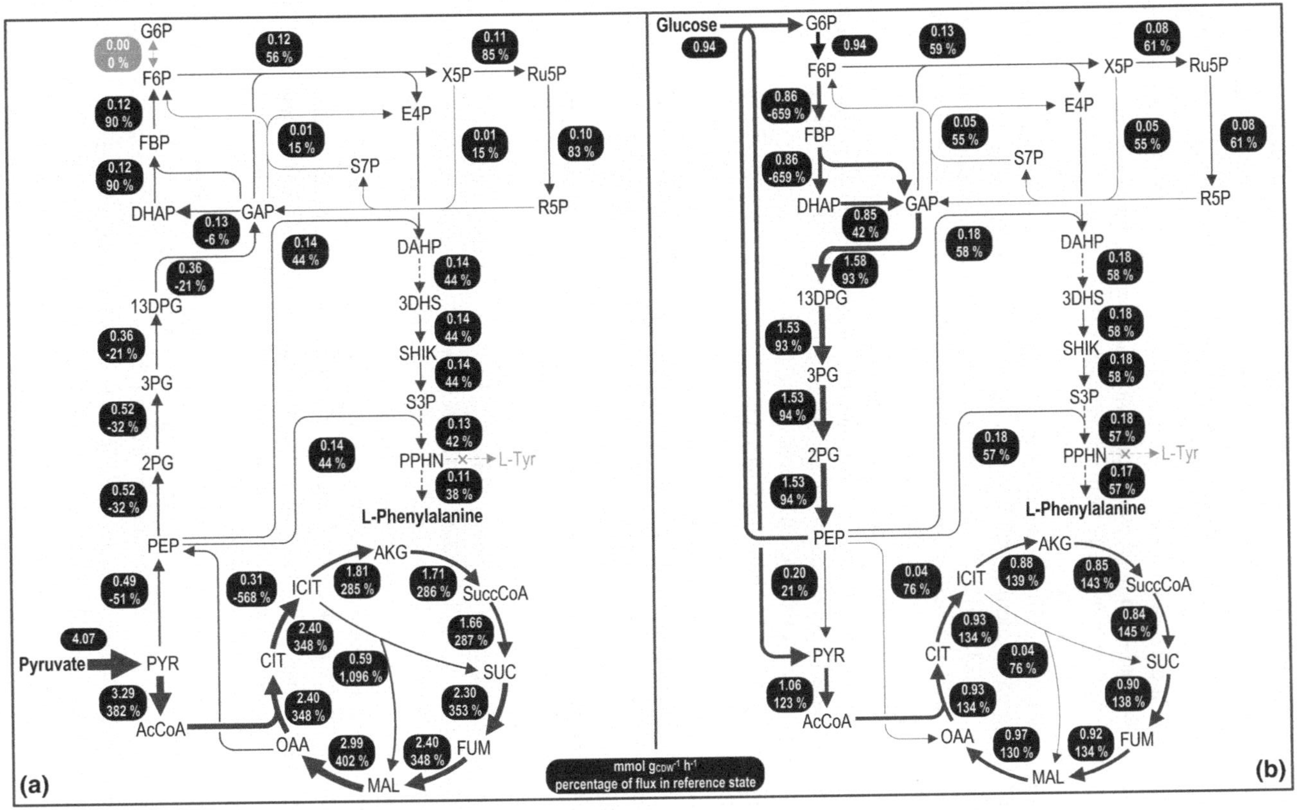

Fig. 11 Intracellular flux distributions (*g*FBA) during perturbation experiments on pyruvate (**a**) and glucose (**b**). Constraints: substrate uptake and L-phenylalanine production during highest substrate supply. *Upper value* absolute flux in mmol g_{CDW}^{-1} h^{-1}, *lower value* percentage of flux to reference state (15 L production process)

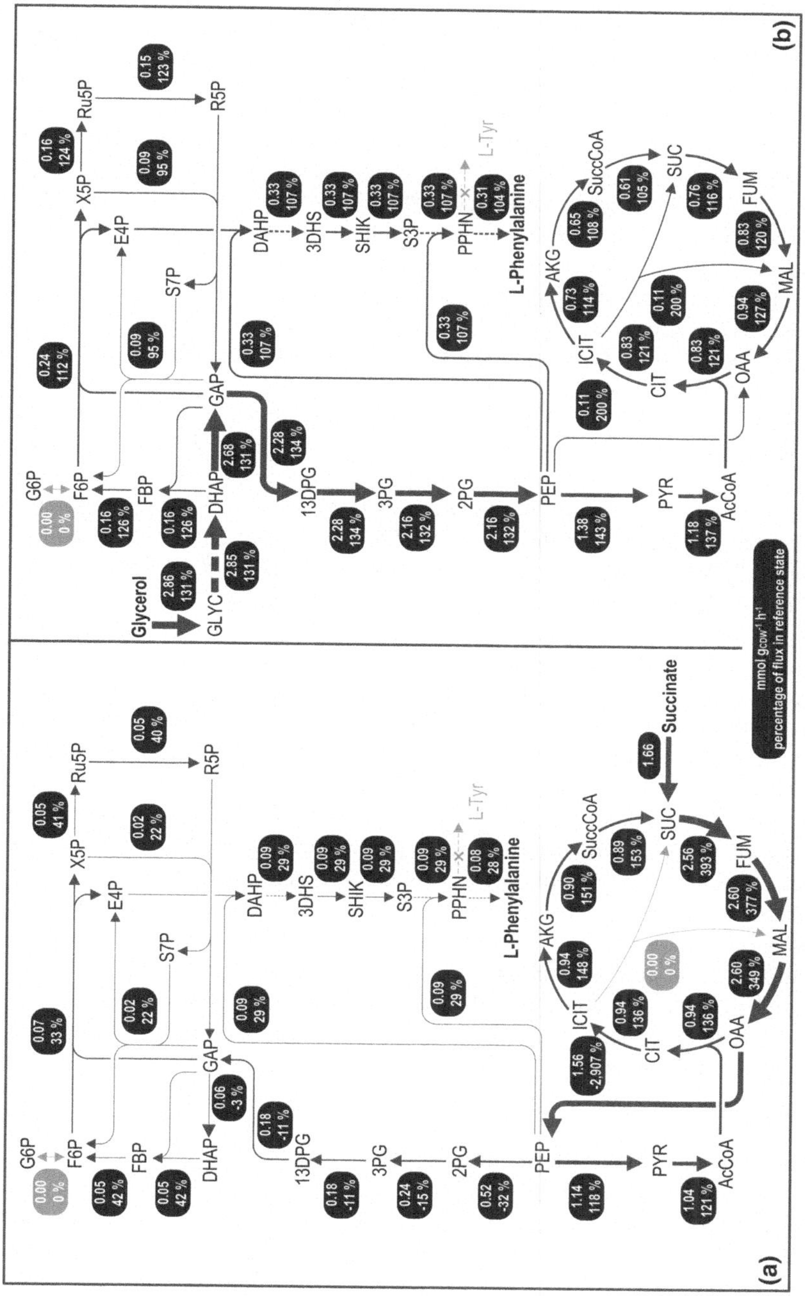

Fig. 12 Intracellular flux distributions (*g*FBA) during perturbation experiments on succinate (**a**) and glycerol (**b**). Constraints: substrate uptake and L-phenylalanine production during highest substrate supply. *Upper value* absolute flux in mmol g_{CDW}^{-1} h^{-1}, *lower value* percentage of flux to reference state (15 L production process)

to DHAP, which served as starting point for gluconeogenetic and glycolytic reactions. A total of 0.16 mmol g_{CDW}^{-1} h^{-1} F6P and 2.16 mmol g_{CDW}^{-1} h^{-1} PEP were generated. E4P as a precursor for L-phenylalanine was obtained by the transformation of F6P and GAP in the pentose phosphate pathway. Next to E4P, PEP was necessary in two enzymatic conversions for the constrained L-phenylalanine production (0.31 mmol g_{CDW}^{-1} h^{-1}). PEP was further used to generate PYR and subsequently AcCoA, fueling the citric acid cycle. This was supported by the direct conversion of PEP to OAA (0.11 mmol g_{CDW}^{-1} h^{-1}).

The aim of perturbation experiments is a various deflection around the metabolic reference state. To illustrate this, the percentages of these perturbation fluxes in relation to reference fluxes of the 15 L-scale process are given in the Figs. 11 and 12. The values of intracellular reference fluxes are depicted in Table 3. The higher uptake of glycerol in the metabolic analysis compared to the process (131 %) increased all fluxes to 107–200 %, including L-phenylalanine production. With glucose as carbon source, the direction of most reactions remained unchanged with reduced activities (55–94 %). Fluxes through the citric acid cycle were increased to ~140 %, and the reactions in upper glycolysis (G6P to DHAP) were activated or reversed with an almost sevenfold activity. Moreover, the usage of PEP and the production of PYR during glucose uptake influenced the PEP-PYR node. The flux distributions with the carbon sources succinate and pyruvate had several characteristics in common. With both entering the metabolism in or near the citric acid cycle, the flux from PEP to F6P was reversed and less pronounced in comparison with the reference (~−20 %). Reduced activities occurred in pentose phosphate pathway and the aromatic amino acid pathway (~30 % or ~40 %). Pyruvate as carbon source caused reversed fluxes between PYR-PEP with halved activity and OAA-PEP with almost sixfold activity. Strong increases of fluxes resulted in citric acid cycle (~300 %) and glyoxylate bypass (1,096 %). However, succinate as carbon source amplified the metabolic fluxes from succinate to oxaloacetate almost by a factor of four. The remaining reactions in citric acid cycle were increased by approximately 40 %, and the glyoxylate bypass was inactive. Succinate also reversed the flux between OAA and PEP with an even higher activity (−2,907 %).

In summary, the applied different carbon sources caused the varying perturbations aimed at. Next to changes in L-phenylalanine production rates, the activities in central metabolism changed strongly and several reactions were reversed depending on the applied carbon source.

5.4.2 Intracellular Metabolite Quantification

For further characterization of cell metabolism during L-phenylalanine production, intracellular metabolome was quantified during parallelized steady-state fed-batch perturbation experiments. Three intracellular samples were taken in each perturbation reactor at a cultivation time of 6, 12, and 18 min to represent intracellular metabolome at steady-state conditions. Sampling was performed using an improved version of the rapid sampling device proposed by [90]. For inactivation of cell

Table 3 Intracellular flux distribution (*g*FBA) in 15 L-scale reference process during parallelized metabolic perturbation experiments

Reaction	Flux
Glycerol metabolism	
Glycerol uptake	2.18
Glycerol kinase	2.18
Glycerol-3-phosphate dehydrogenase	2.18
Glycolysis	
Fructose-bisphosphatase	0.13
Fructose-bisphosphate aldolase	0.13
Triose-phosphate isomerase	2.05
Glyceraldehyde-3-phosphate dehydrogenase	1.70
Phosphoglycerate kinase	1.70
Phosphoglycerate mutase	1.64
Enolase	1.64
Pyruvate kinase	0.97
Citric acid cycle	
Pyruvate dehydrogenase	0.86
Citrate synthase	0.69
Aconitase	0.69
Isocitrate dehydrogenase	0.64
α-Ketogluterate dehydrogenase	0.60
Succinyl-CoA synthetase	0.58
Succinate dehydrogenase	0.65
Fumarase	0.69
Malate dehydrogenase	0.74
Pentose phosphate pathway	
Ribulose-5-phosphate epimerase	0.13
Ribose-5-phosphate isomerase	0.12
Transketolase I	0.09
Transketolase II	0.22
Transaldolase	0.09
Biosynthesis of aromatic amino acids	
3-deoxy-D-arabino-heptulosonate-7-phosphate synthetase	0.31
3-Dehydroquinate synthase	0.31
3-Dehydroquinate dehydratase	0.31
Shikimate dehydrogenase	0.31
Shikimate kinase	0.31
3-phosphoshikimate 1-carboxyvinyltransferase	0.31
Chorismate synthase	0.31
Chorismate mutase	0.31
Prephenate dehydratase	0.31

(continued)

Table 3 (continued)

Reaction	Flux
Phenylalanine transaminase	0.31
L-Phenylalanine exchange	0.30
Anaplerotic reactions	
Isocitrate lyase	0.05
Malate synthase	0.05
Phosphoenolpyruvate carboxylase	0.05

Applied constraints: Glycerol uptake and L-phenylalanine production. Flux values given in mmol g_{CDW}^{-1} h^{-1}

metabolism, a cold (−70 °C) methanol solution (60 %, 30 mM TEA) was used (quenching ratio culture broth:quenching agent 1:3). Extraction was performed in hot buffered water (95 °C, 30 mM TEA). For correct quantification, an intracellular U-^{13}C cell extract was added prior to extraction. Finally, quantification was performed using (U)HPLC–mass spectrometry (TSQ Vantage, Thermo Scientific, USA) in combination with a high-performance method for intracellular metabolite quantification [114]. For quantitative correction of intracellular metabolites in culture broth, extracellular samples were taken, prepared analog to intracellular samples and also quantified by (U)HPLC-MS.

In Fig. 13, intracellular metabolite concentrations of fructose-1,6-bisphosphate (FBP), phosphoenolpyruvate (PEP), sedoheptulose-7-phosphate (S7P), and shikimate-3-phosphate (S3P) are shown exemplarily. For all shown intracellular metabolites, a targeted deflection around reference state (15 L production process) was successfully detected during all perturbation experiments. Further, clear variations in metabolite concentrations were obtained as a function of the applied carbon source. By applying the rationally designed feeding profile for perturbation experiments, a good accordance between reference state (15 L production process on glycerol) and perturbation experiments on glycerol was obtained. Comparable to extracellular fluxes, intracellular metabolite concentrations of the reference state were between steady-state B and C of the perturbation experiment on glycerol. This accordance between process reactor and perturbation reactor underlines applicability of 'Rapid Media Transition' for parallelized perturbation experiments.

In case of the upper glycolysis intermediate FBP, a higher metabolome concentration was measured when substrate uptake was increased. This result was obtained for all applied carbon sources. FBP concentration was deflected within a range of ~300 μM.

For PEP, nearly constant intracellular concentrations of ~1.0 and ~1.25 mM were measured during steady states on pyruvate and succinate. The increasing substrate uptake during metabolic steady states on glucose and glycerol resulted in a clear decrease of PEP concentration.

Decrease of PEP concentration during increasing glucose supply correlates well with literature. Xu et al. [140] demonstrated that *E. coli* quickly and almost completely turns off PEP consumption when glucose availability is limited. By reducing

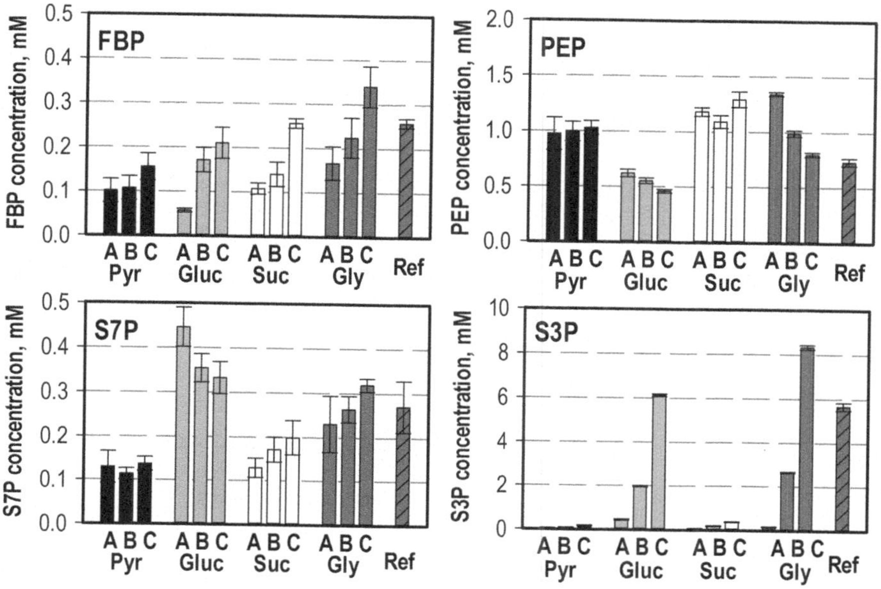

Fig. 13 Intracellular metabolite concentrations of fructose-1,6-bisphosphate (FBP), phosphoenolpyruvate (PEP), sedoheptulose-7-phosphate (S7P), and shikimate-3-phosphate (S3P) during 15 L production process (Ref) and parallelized fed-batch perturbation experiments on pyruvate (Pyr), glucose (Gluc), succinate (Suc), and glycerol (Gly). *A*, *B*, *C* time of sampling (6, 12, and 18 min) during perturbation experiments according to feed profile (Fig. 8)

PEP consumption at low substrate availability, high uptake rates are ensured at sudden substrate excess. This active regulation of glucose uptake is illustrated in Fig. 14. Further, negative regulation of phosphofructokinase (pfk) by PEP was measured during experiments on glucose. In vivo kinetic is shown in Fig. 14. This result is also in good accordance with in vivo experiments of [106] and in vitro results of [141].

Since two molecules of PEP are metabolized for L-phenylalanine synthesis, decrease of PEP concentration during perturbation experiments on glycerol may be attributed to this raising PEP demand during stepwise increased L-phenylalanine synthesis. Within all experiments, PEP concentration was deflected within a range of ~800 μM.

Concentration of S7P, as intermediate of pentose phosphate pathway, was constant (~120 μM) during experiments on pyruvate, whereas usage of glucose led to high S7P concentration (up to ~450 μM). A further increase of glucose supply decreased S7P availability. In contrast, S7P concentration was increased by higher carbon supply on succinate and glycerol. During all parallelized perturbation experiments, S7P concentration was deflected within a range of ~330 μM.

The highest deflection of intracellular metabolite concentration was measured for the amino acid pathway intermediate S3P. Whereas only low concentrations (<200 μM) were detectable on pyruvate and succinate as carbon source, S3P

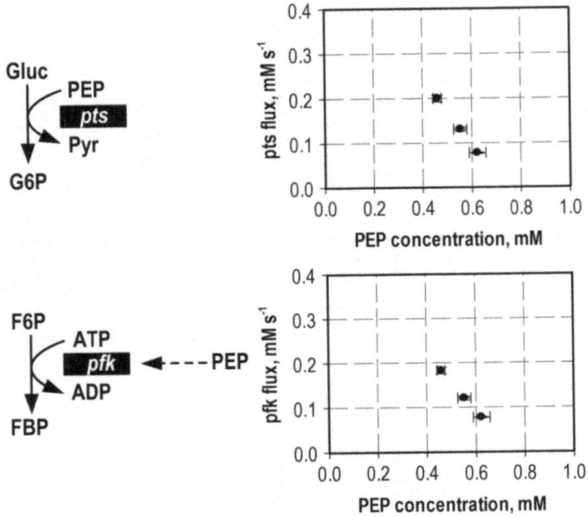

Fig. 14 Fluxes of phosphotransferase system (pts) and phosphofructokinase (pfk) are plotted against the concentration of phosphoenolpyruvate (PEP) in experiments with glucose as carbon source. Flux variance was estimated by calculation of multiple FVAs under systematic consideration of the errors present in extracellular flux determination

concentration was significantly increased on glucose and glycerol. For these two carbon sources, concentrations of up to 8 mM were quantified. This example clearly emphasizes the differences between the applied carbon sources when used for L-phenylalanine synthesis. The carbon sources of upper glycolysis benefit metabolite supply in the aromatic amino acid pathway, whereas for carbon sources of lower glycolysis or TCA, carbon supply has to be realized by gluconeogenesis and therefore is not as effective as by glucose or glycerol.

In summary, by application of different carbon sources, targeted deflection of metabolome concentration was successfully achieved. The magnitude of deflection was highly dependent on the used carbon source and the realized substrate uptake rate. For all metabolites, perturbation results were in good accordance with reference state and literature.

6 Conclusions

The aim of metabolic engineering in industrial biotechnology is the optimization of microbial production strains to improve process performance and product yields. For targeted changes in metabolism, it is necessary to have a comprehensive insight into the actual metabolic activities of the producer cells in the production reactor. We gave a broad overview over various methods that are applied to gain this kind of information. These include several kinds of perturbation experiments, several approaches for theoretical and experimental metabolic flux analyses, and moreover for the exact quantification of intracellular metabolites. These fields are closely related to each other and therefore cannot be dealt with as completely separate topics.

Furthermore, we presented an extended experimental approach that enables parallel data generation for the characterization of cell metabolism during an industrial fed-batch process with minimal disturbance of the process. Therefore, the method of 'Rapid Media Transition' [19] was used. To this end, culture broth is withdrawn from the process; the cells are separated rapidly from the medium and are transferred directly to four parallelized analytical stirred-tank bioreactors. This setup allows four parallel perturbation experiments with cells of the same origin and proteomic state. Therefore, the generation of a huge data set about this particular metabolic origin is realized. This point is of central importance because production processes lead to fast changing conditions. Therefore, comparability of consecutive applied perturbation experiments would not be guaranteed.

Samples for the analysis of extra- and intracellular metabolites were studied and revealed a successful deflection of the metabolic state in comparison with the reference process. The data gained about extra- and intracellular fluxes as well as intracellular metabolite concentrations may, for example, be used for a metabolic control analysis, which is able to indicate bottlenecks in production metabolism.

The method of parallelized perturbation experiments is especially capable of analyzing production processes in detail as these experiments may be executed repeatedly during one continuing fed-batch process. By doing so, metabolic changes that occur in the cells during the process can be uncovered. This might, for example, give evidence why the production phase ends at a certain time and therefore suggest specific genetic changes to improve the process.

Acknowledgments Support of the German Research Foundation DFG (Grant numbers WE 2715/11-1, SP 503/7-1) is gratefully acknowledged.

References

1. Feist AM, Zielinski DC, Orth JD, Schellenberger J, Herrgard MJ, Palsson BØ (2010) Model-driven evaluation of the production potential for growth-coupled products of *Escherichia coli*. Metab Eng 12(3):173–186
2. Wiechert W, Noack S (2011) Mechanistic pathway modeling for industrial biotechnology: challenging but worthwhile. Curr Opin Biotechnol 22(5):604–610
3. Almquist J, Cvijovic M, Hatzimanikatis V, Nielsen J, Jirstrand M (2014) Kinetic models in industrial biotechnology—improving cell factory performance. Metab Eng 24:38–60
4. Oberhardt MA, Palsson BØ, Papin JA (2009) Applications of genome-scale metabolic reconstructions. Mol Sys Biol 5:320
5. Stephanopoulos GN, Aristidou AA, Nielsen J (1998) Metabolic engineering—principles and methodologies. Academic Press, San Diego

6. Oldiges M, Takors R (2005) Applying metabolic profiling techniques for stimulus-response experiments: chances and pitfalls. Adv Biochem Eng Biotechnol 92:173–196
7. Magnus JB, Hollwedel D, Oldiges M, Takors R (2006) Monitoring and modeling of the reaction dynamics in the valine/leucine synthesis pathway in *Corynebacterium glutamicum*. Biotechnol Prog 22:1071–1083
8. Visser D, van Zuylen GA, van Dam JC, Eman MR, Pröll A, Ras C, Wu L, van Gulik WM, Heijnen JJ (2004) Analysis of in vivo kinetics of glycolysis in aerobic *Saccharomyces cerevisiae* by application of glucose and ethanol pulses. Biotechnol Bioeng 88(2):157–167
9. Taymaz-Nikerel H, van Gulik WM, Heijnen JJ (2011) *Escherichia coli* responds with a rapid and large change in growth rate upon a shift from glucose-limited to glucose-excess conditions. Metab Eng 13(3):307–318
10. Mashego MR, van Gulik WM, Heijnen JJ (2007) Metabolome dynamic responses of *Saccharomyces cerevisiae* to simultaneous rapid perturbations in external electron acceptor and electron donor. FEMS Yeast Res 7(1):48–66
11. Wu L, Mashego MR, Pröll AM, Vinke JL, Ras C, van Dam J, van Winden WA, van Gulik WM, Heijnen JJ (2006) In vivo kinetics of primary metabolism in *Saccharomyces cerevisiae* studied through prolonged chemostat cultivation. Metab Eng 8(2):160–171
12. Nasution U, van Gulik WM, Pröll A, van Winden WA, Heijnen JJ (2006) Generating short-term kinetic responses of primary metabolism of *Penicillium chrysogenum* through glucose perturbation in the bioscope mini reactor. Metab Eng 8(5):395–405
13. Chassagnole C, Noisommit-Rizzi N, Schmid JW, Mauch K, Reuss M (2002) Dynamic modeling of the central carbon metabolism of *Escherichia coli*. Biotechnol Bioeng 79(1):53–73
14. Aboka FO, Yang H, de Jonge LP, Kerste R, van Winden WA, van Gulik WM, Hoogendijk R, Oudshoorn A, Heijnen JJ (2006) Characterization of an experimental miniature bioreactor for cellular perturbation studies. Biotechnol Bioeng 95(6):1032–1042
15. Visser D, van Zuylen GA, van Dam JC, Oudshoorn A, Eman MR, Ras C, van Gulik WM, Frank J, van Dedem GWK, Heijnen JJ (2002) Rapid sampling for analysis of in vivo kinetics using the BioScope: a system for continuous-pulse experiments. Biotechnol Bioeng 79 (6):674–681
16. Mashego MR, van Gulik WM, Vinke JL, Visser D, Heijnen JJ (2006) In vivo kinetics with rapid perturbation experiments in *Saccharomyces cerevisiae* using a second-generation BioScope. Metab Eng 8(4):370–383
17. Kresnowati MTAP, van Winden WA, van Gulik WM, Heijnen JJ (2008) Dynamic in vivo metabolome response of *Saccharomyces cerevisiae* to a stepwise perturbation of the ATP requirement for benzoate export. Biotechnol Bioeng 99(2):421–441
18. Nasution U, van Gulik WM, Ras C, Pröll A, Heijnen JJ (2008) A metabolome study of the steady-state relation between central metabolism, amino acid biosynthesis and penicillin production in *Penicillium chrysogenum*. Metab Eng 10(1):10–23
19. Link H, Anselment B, Weuster-Botz D (2010) 'Rapid media transition': an experimental approach for steady-state analysis of metabolic pathways. Biotechnol Prog 26(1):1–10
20. Sauer U (2006) Metabolic networks in motion: ^{13}C-based flux analysis. Mol Sys Biol 2:68–10
21. Quek LE, Wittmann C, Nielsen LK, Krömer JO (2009) OpenFLUX: efficient modelling software for ^{13}C-based metabolic flux analysis. Microb Cell Fact 8:25
22. Zamboni N, Fischer E, Sauer U (2005) FiatFlux—a software for metabolic flux analysis from 13C-glucose experiments. BMC Bioinformatics 6:209
23. Weitzel M, Nöh K, Dalman T, Niedenführ S, Stute B, Wiechert W (2013) 13CFLUX2—high-performance software suite for ^{13}C-metabolic flux analysis. Bioinformatics 29(1):143–145
24. Wasylenko TM, Stephanopoulos G (2015) Metabolomic and ^{13}C-metabolic flux analysis of a xylose-consuming Saccharomyces cerevisiae strain expressing xylose isomerase. Biotech Bioeng 112(3):470–483
25. Antoniewicz MR (2015) Methods and advances in metabolic flux analysis: a mini-review. J Ind Microbiol Biotechnol 42:317–325
26. Antoniewicz MR (2013) ^{13}C metabolic flux analysis: optimal design of isotopic labeling experiments. Curr Opin Biotech 24:1116–1121

27. Leighty RW, Antoniewicz MR (2011) Dynamic metabolic flux analysis (DMFA): a framework for determining fluxes at metabolic non-steady state. Metab Eng 13:745–755
28. Buescher JM, Antoniewicz MR, Boros LG, Burgess SC, Brunengraber H, Clish CB, DeBerardinis RJ, Feron O, Frezza C, Ghesquiere B, Gottlieb E, Hiller K, Jones RG, Kamphorst JJ, Kibbey RG, Kimmelman AC, Locasale JW, Lunt SY, Maddocks ODK, Malloy C, Metallo CM, Meuillet EJ, Munger J, Nöh K, Rabinowitz JD, Ralser M, Sauer U, Stephanopoulos G, St-Pierre J, Tennant DA, Wittmann C, Vander Heiden MG, Vazquez A, Vousden K, Young JD, Zamboni N, Fendt SM (2015) A roadmap for interpreting ^{13}C metabolite labeling patterns from cells. Curr Opin Biotech 34:189–201
29. Wiechert W (2001) ^{13}C metabolic flux analysis. Metab Eng 3:195–206
30. Nöh K, Grönke K, Luo B, Takors R, Oldiges M, Wiechert W (2007) Metabolic flux analysis at ultra short time scale: isotopically non-stationary ^{13}C labeling experiments. J Biotech 129 (2):249–267
31. Schaub J, Mauch K, Reuss M (2008) Metabolic flux analysis in *Escherichia coli* by integrating isotopic dynamic and isotopic stationary ^{13}C labeling data. Biotechnol Bioeng 99 (5):1170–85
32. Zamboni N, Fendt SM, Rühl M, Sauer U (2009) ^{13}C-based metabolic flux analysis. Nat Protoc 4(6):878–92
33. Noack S, Nöh K, Moch M, Oldiges M, Wiechert W (2011) Stationary versus non-stationary ^{13}C-MFA: a comparison using a consistent dataset. J Biotech 154:179–190
34. Nöh K, Wiechert W (2011) The benefits of being transient: isotope-based metabolic flux analysis at the short time scale. Appl Microbiol Biotechnol 91:1247–1265
35. Wiechert W, Nöh K (2013) Isotopically non-stationary metabolic flux analysis: complex yet highly informative. Mol Sys Biol 24(6):979–86
36. Nöh K, Wiechert W (2006) Experimental design principles for isotopically instationary ^{13}C labeling experiments. Biotechnol Bioeng 94(2):234–51
37. Young JD, Walther JL, Antoniewicz MR, Yoo H, Stephanopoulos G (2008) An elementary metabolite unit (EMU) based method of isotopically nonstationary flux analysis. Biotech Bioeng 99(3):686–699
38. Reed JL, Palsson BØ (2003) Thirteen years of building constraint-based in silico models of *Escherichia coli*. J Bacteriol 185(9):2692–2699
39. Bordbar A, Monk JM, King ZA, Palsson BØ (2014) Constraint-based models predict metabolic and associated cellular functions. Nat Rev Genet 15(2):107–120
40. McCloskey D, Palsson BØ, Feist AM (2013) Basic and applied uses of genome-scale metabolic network reconstructions of *Escherichia coli*. Mol Sys Biol 9:661
41. Orth JD, Palsson BØ (2010) What is flux balance analysis. Nat Biotechnol 28(3):245–248
42. Feist AM, Palsson BØ (2008) The growing scope of applications of genome-scale metabolic reconstructions using *Escherichia coli*. Nat Biotechnol 26:659–667
43. Meadows AL, Karnik R, Lam H, Forestell S, Snedecor B (2010) Application of dynamic flux balance analysis to an industrial *Escherichia coli* fermentation. Metab Eng 12:150–160
44. Weiner M, Tröndle J, Albermann C, Sprenger GA, Weuster-Botz D (2014) Improvement of constraint-based flux estimation during L-phenylalanine production with *Escherichia coli* using targeted knock-out mutants. Biotechnol Bioeng 111:1406–1416
45. Beard DA, Liang SD, Qian H (2002) Energy balance for analysis of complex metabolic networks. Biophys J 83:79–86
46. Price ND, Famili I, Beard DA, Palsson BØ (2002) Extreme pathways and Kirchhoff's second law. Biophys J 83:2879–2882
47. Reed J (2012) Shrinking the metabolic solution space using experimental datasets. Biophys J 8(8):e1002662
48. Majewski RA, Domach MM (1990) Simple constrained optimization view of acetate overflow in *E. coli*. Biotechnol Bioeng 35:732–738
49. Reed JL, Vo TD, Schilling CH, Palsson BØ (2003) An expanded genome-scale model of *Escherichia coli* K-12 (*i*JR904 GSM/GPR). Genome Biol 4(9):R54

50. Blattner FR, Plunkett G, Bloch CA, Perna NT, Burland V, Riley M, Collado-Vides J, Glasner JD, Rode CK, Mayhew GF, Gregor J, Davis NW, Kirkpatrick HA, Goeden MA, Rose DJ, Mau B, Shao Y (1997) The complete genome sequence of *Escherichia coli* K-12. Science 277:1453–1474
51. Hayashi K, Morooka N, Yamamoto Y, Fujita K, Isono K, Choi S, Ohtsubo E, Baba T, Wanner BL, Mori H, Horiuchi T (2006) Highly accurate genome sequences of *Escherichia coli* K-12 strains MG1655 and W3110. Mol Syst Biol 2(1).
52. Feist AM, Henry CS, Reed JL, Krummenacker M, Joyce AR, Karp PD, Broadbelt LJ, Hatzimanikatis V, Palsson BØ (2007) A genome-scale metabolic reconstruction for *Escherichia coli* K-12 MG1655 that accounts for 1260 ORFs and thermodynamic information. Mol Syst Biol 3:121
53. Orth JD, Conrad TM, Na J, Lerman JA, Nam H, Feist AM, Palsson BØ (2011) A comprehensive genome-scale reconstruction of *Escherichia coli* metabolism–2011. Mol Syst Biol 7:535
54. Förster J, Famili I, Fu P, Palsson BØ, Nielsen J (2003) Genome-scale reconstruction of the *Saccharomyces cerevisiae*. Genome Res 13:244–253
55. Nookaew I, Jewett MC, Meechai A, Thammarongtham C, Laoteng K, Cheevadhanarak S, Nielsen J, Bhumiratana S (2008) Genome-scale reconstruction of the *Saccharomyces cerevisiae*. Genome Res 2:71
56. Nookaew I, Olivares-Hernández R, Bhumiratana S, Nielsen J (2011) Genome-scale metabolic models of *Saccharomyces cerevisiae*. Methods Mol Biol 759:445–463
57. Dias O, Pereira R, Gombert AK, Ferreira EC, Rocha I (2014) *i*OD907, the first genome-scale metabolic model for the milk yeast *Kluyveromyces lactis*. Biotechnol J 9(6):776–790
58. McAnulty MJ, Yen JY, Freedman BG, Senger RS (2012) Genome-scale modeling using flux ratio constraints to enable metabolic engineering of clostridial metabolism in silico. BMC Sys Biol 6:42
59. Goyal N, Widiastuti H, Karimi IA, Zhou Z (2014) A genome-scale metabolic model of *Methanococcus maripaludis* S2 for CO_2 capture and conversion to methane. Mol Biosyst 10 (5):1043–54
60. Wu X, Wang X, Lu W (2014) Genome-scale reconstruction of a metabolic network for *Gluconobacter oxydans* 621H. Biosystems 117:10–14
61. Becker SA, Feist AM, Mo ML, Hannum G, Palsson BØ, Herrgard MJ (2007) Quantitative prediction of cellular metabolism with constraint-based models: the COBRA Toolbox. Nat Protoc 2(3):727–738
62. Schellenberger J, Que R, Fleming RMT, Thiele I, Orth JD, Feist AM, Zielinski DC, Bordbar A, Lewis NE, Rahmanian S, Kang J, Hyduke DR, Palsson BØ (2011) Quantitative prediction of cellular metabolism with constraint-based models: the COBRA Toolbox v2.0. Nat Protoc 6(9):1290–1307
63. Hucka M, Finney A, Sauro HM, Bolouri H, Doyle JC, Kitano H, Arkin AP, Bornstein BJ, Bray D, Cornish-Bowden A, Cuellar AA, Dronov S, Gilles ED, Ginkel M, Gor V, Goryanin II, Hedley WJ, Hodgman TC, Hofmeyr JH, Hunter PJ, Juty NS, Kasberger JL, Kremling A, Kummer U, Le Novère N, Loew LM, Lucio D, Mendes P, Minch E, Mjolsness ED, Nakayama Y, Nelson MR, Nielsen PF, Sakurada T, Schaff JC, Shapiro BE, Shimizu TS, Spence HD, Stelling J, Takahashi K, Tomita M, Wagner J, Wang J, Forum SBML (2003) The systems biology markup language (SBML): a medium for representation and exchange of biochemical network models. Bioinformatics 19(4):524–31
64. Schellenberger J, Park JO, Conrad TM, Palsson BØ (2010) BiGG: a biochemical genetic and genomic knowledgebase of large scale metabolic reconstructions. BMC Bioinformatics 11:213
65. Schuetz R, Kuepfer L, Sauer U (2007) Systematic evaluation of objective functions for predicting intracellular fluxes in *Escherichia coli*. Mol Syst Biol 3:119
66. Schilling CH, Edwards JS, Letscher D, Palsson BØ (2001) Combining pathway analysis with flux balance analysis for the comprehensive study of metabolic systems. Biotechnol Bioeng 71(4):286–310

67. Edwards JS, Palsson BØ (2000) The *Escherichia coli* MG1655 in silico metabolic genotype: its definition, characteristics, and capabilities. Proc Natl Acad Sci USA 97(10):5528–5533
68. Mahadevan R, Schilling CH (2003) The effects of alternate optimal solutions in constraint-based genome-scale metabolic models. Metab Eng 5:264–276
69. Schellenberger J, Lewis NE, Palsson BØ (2011) Elimination of thermodynamically infeasible loops in steady-state metabolic models. Biophys J 100(3):544–553
70. Smallbone K, Simeonidis E (2009) Flux balance analysis: A geometric perspective. J Theor Biol 258(2):311–315
71. Mahadevan R, Schilling CH (2003) The effects of alternate optimal solutions in constraint-based genome-scale metabolic models. Metab Eng 5(4):264–276
72. Kümmel A, Panke S, Heinemann M (2006) Putative regulatory sites unraveled by network-embedded thermodynamic analysis of metabolome data. Mol Syst Biol 2:1–10
73. Henry CS, Broadbelt LJ, Hatzimanikatis V (2007) Thermodynamics-based metabolic flux analysis. Biophys J 92(5):1792–1805
74. Fleming RMT, Thiele I, Provan G, Nasheuer HP (2010) Integrated stoichiometric, thermodynamic and kinetic modelling of steady-state metabolism. J Theor Biol 264 (3):683–692
75. Müller AC, Bockmayr A (2013) Fast thermodynamically constrained flux variability analysis. Bioinformatics 29(7):903–909
76. Holzhütter H (2004) The principle of flux minimization and its application to estimate stationary fluxes in metabolic networks. Eur J Biochem 271:2905–2922
77. Goodacre R, Vaidyanathan S, Dunn WB, Harrigan GG, Kell DB (2004) Metabolomics by numbers: acquiring and understanding global metabolite data. Trends Biotechnol 22(5):245–52
78. Rabinowitz JD (2007) Cellular metabolomics of *Escherichia coli*. Expert Rev Proteomics 4 (2):187–198
79. Gancedo JM, Gancedo C (1973) Concentrations of intermediary metabolites in yeast. Biochemie 55:205–211
80. Lowry OH, Carter J, Ward JB, Glaser L (1971) The effect of carbon and nitrogen sources on the level of metabolic intermediates in *Escherichia coli*. J Biol Chem 246:6511–6521
81. Tweeddale H, Notley-McRobb L, Ferenci T (1998) Effect of slow growth on metabolism of *E. coli* as revealed by global metabolite pool ("Metabolome") analysis. J Bacteriol 180 (19):5109–5116
82. Cotten C, Reed JL (2013) Mechanistic analysis of multi-omics datasets to generate kinetic parameters for constraint-based metabolic models. BMC Bioinformatics 14:32
83. Mashego MR, Rumbold K, De Mey M, Vandamme E, Soetaert W, Heijnen JJ (2007) Microbial Metabolomics: Past, present and future methodologies. Biotechnol Lett 29(1):1–16
84. Chubukov V, Gerosa L, Kochanowski K, Sauer U (2014) Coordination of microbial metabolism. Nat Rev Microbiol 12(5):327–40
85. Zenobi R (2013) Single-cell metabolomics: analytical and biological perspectives. Science 342(6163):1243–1259
86. de Koning W, van Dam K (1992) A method for the determination of changes of glycolytic metabolites in yeast on a subsecond time scale using extraction at neutral PH. Anal Biochem 204:118–123
87. Schadel F, Franco-Lara E (2009) Rapid sampling devices for metabolic engineering applications. Appl Microbiol Biotechnol 83(2):199–208
88. Lange HC, Eman M, van Zuijlen G, Visser D, van Dam J, Frank J, Teixeira de Mattos MJ, Heijnen JJ (2001) Improved rapid sampling for in vivo kinetics of intracellular metabolites in *Saccharomyces Cerevisiae*. Biotechnol Bioeng 75(4):406–415
89. Schaub J, Schiesling C, Reuss M, Dauner M (2006) Integrated sampling procedure for metabolome analysis. Biotechnol Prog 22:1434–1442
90. Hiller J, Franco-Lara E, Papaioannou V, Weuster-Botz D (2007) Fast sampling and quenching procedures for microbial metabolic profiling. Biotechnol Lett 29(8):1161–1167
91. McCloskey D, Gangoiti JA, King ZA, Naviaux RK, Barshop BA, Palsson BØ, Feist AM (2014) A model-driven quantitative metabolomics analysis of aerobic and anaerobic

metabolism in *E. Coli* K-12 Mg1655 that is biochemically and thermodynamically consistent. Biotechnol Bioeng 111(4):803–815
92. Larsson G, Törnkvist M (1996) Rapid sampling cell inactivation and evaluation of low extracellular glucose concentrations during fed-batch cultivations. J Biotech 49:69–82
93. Weuster-Botz D (1997) Sampling tube device for monitoring intracellular metabolite dynamics. Anal Chem 246:225–233
94. Schaefer U, Boos W, Takors R, Weuster-Botz D (1999) Automated sampling device for monitoring intracellular metabolite dynamics. Anal Chem 270:88–96
95. Bolten CJ, Kiefer P, Letisse F, Portais JC, Wittmann C (2007) Sampling for metabolome analysis of microorganisms. Anal Chem 79(10):3843–3849
96. Link H, Buescher JM, Sauer U (2012) Targeted and quantitative metabolomics in bacteria. In: Wipat A, Harwood C (eds) Systems biology of bacteria. Elsevier, Amsterdam
97. da Luz JA, Hans E, Zeng AP (2014) Automated fast filtration and on-filter quenching improve the intracellular metabolite analysis of microorganisms. Eng Life Sci 14:135–142
98. Winder CL, Dunn WB, Schuler S, Broadhurst D, Jarvis R, Stephens GM, Goodacre R (2008) Global metabolic profiling of *Escherichia Coli* cultures: an evaluation of methods for quenching and extraction of intracellular metabolites. Anal Chem 80(8):2939–2948
99. Link H, Anselment B, Weuster-Botz D (2008) Leakage of adenylates during cold methanol/glycerol quenching of *Escherichia Coli*. Metabolomics 4(3):240–247
100. Taymaz-Nikerel H, de Mey M, Ras C, ten Pierick A, Seifar RM, van Dam JC, Heijnen JJ, van Gulik WM (2009) Development and application of a differential method for reliable metabolome analysis in *Escherichia Coli*. Anal Biochem 386(1):9–19
101. van Gulik WM (2010) Fast sampling for quantitative microbial metabolomics. Curr Opin Biotechnol 21(1):27–34
102. Rabinowitz JD, Kimball E (2007) Acidic acetonitrile for cellular metabolome extraction from *Escherichia Coli*. Anal Chem 79:6167–6173
103. Paczia N, Nilgen A, Lehmann T, Gatgens J, Wiechert W, Noack S (2012) Extensive exometabolome analysis reveals extended overflow metabolism in various microorganisms. Microb Cell Fact 11(1):122
104. Annesley TM (2003) Ion suppression in mass spectrometry. Clin Chem 49(7):1041–1044
105. Jaki BU, Franzblau SG, Cho SH, Pauli GF (2006) Development of an extraction method for mycobacterial metabolome analysis. J Pharm Biomed Anal 41(1):196–200
106. Link H (2009) 'Rapid media transition' for metabolic control analysis of fed-batch fermentation processes. Ph.D thesis, Technische Universität München.
107. Zimmermann M, Sauer U, Zamboni N (2014) Quantification and mass isotopomer profiling of alpha-keto acids in central carbon metabolism. Anal Chem 86(6):3232–7
108. Weiner M, Tröndle J, Schmideder A, Binder K, Albermann C, Sprenger GA, Weuster-Botz D (2015) Parallelized small-scale production of uniformly ^{13}C-labeled cell-extract for quantitative metabolome analysis. Anal Biochem 478:134–140
109. Wu L, Mashego MR, van Dam JC, Proell AM, Vinke JL, Ras C, van Winden WA, van Gulik WM, Heijnen JJ (2005) Quantitative analysis of the microbial metabolome by isotope dilution mass spectrometry using uniformly C-13-labeled cell extracts as internal standards. Anal Biochem 336(2):164–171
110. Bochner BR, Ames BN (1982) Complete analysis of cellular nucleotides by two-dimensional thin layer chromatography. J Biol Chem 257(16):9759–9769
111. Dunn WB (2011) Mass spectrometry in systems biology an introduction. Methods Enzymol 500:15–35
112. Garcia DE, Baidoo EE, Benke PI, Pingitore F, Tang YJ, Villa S, Keasling JD (2008) Separation and mass spectrometry in microbial metabolomics. Curr Opin Microbiol 11 (3):233–239
113. Buescher JM, Czernik D, Ewald JC, Sauer U, Zamboni N (2009) Cross-platform comparison of methods for quantitative metabolomics of primary metabolism. Anal Chem 81(6):2135–2143

114. Buescher JM, Moco S, Sauer U, Zamboni N (2010) Ultrahigh performance liquid chromatography-tandem mass spectrometry method for fast and robust quantification of anionic and aromatic metabolites. Anal Chem 82(11):4403–4412
115. Bingol K, Bruschweiler R (2014) Multidimensional approaches to NMR-Based metabolomics. Anal Chem 86(1):47–57
116. Fuhrer T, Heer D, Begemann B, Zamboni N (2011) High-throughput, accurate mass metabolome profiling of cellular extracts by flow injection-time-of-flight mass spectrometry. Anal Chem 83(18):7074–7080
117. Hsu CC, ElNaggar MS, Peng Y, Fang J, Sanchez LM, Mascuch SJ, Moller KA, Alazzeh EK, Pikula J, Quinn RA, Zeng Y, Wolfe BE, Dutton RJ, Gerwick L, Zhang L, Liu X, Mansson M, Dorrestein PC (2013) Real-time metabolomics on living microorganisms using ambient electrospray ionization flow-probe. Anal Chem 85(15):7014–8
118. Bongaerts J, Kramer M, Muller U, Raeven L, Wubbolts M (2001) Metabolic engineering for microbial production of aromatic amino acids and derived compounds. Metabol Eng 3 (4):289–300
119. Sprenger GA (2007) From scratch to value: engineering *Escherichia Coli* wild type cells to the production of L-phenylalanine and other fine chemicals derived from chorismate. Appl Microbiol Biotechnol 75(4):739–749
120. Choi YJ, Tribe DE (1982) Continuous production of phenylalanine using an *Escherichia-Coli* regulatory mutant. Biotechnol Lett 4(4):223–228
121. Forberg C, Haggstrom L (1988) Phenylalanine production from a rec *Escherichia-Coli* strain in fed-batch culture. J Biotechnol 8(4):291–300
122. Park NH, Rogers PL (1986) L-phenylalanine production in continuous culture using a hyperproducing mutant of *Escherichia-Coli*-K-12. Chem Eng Commun 45(1–6):185–196
123. Ozaki A, Katsumata R, Oka T, Furuya A (1985) Cloning of the genes concerned in phenylalanine biosynthesis in *Corynebacterium Glutamicum* and its application to breeding of a phenylalanine producing strain. Agric Biol Chem 49(10):2925–2930
124. Ikeda M (2006) Towards bacterial strains overproducing L-tryptophan and other aromatics by metabolic engineering. Appl Microbiol Biotechnol 69(6):615–626
125. Backman K, O'Connor MJ, Maruya A, Rudd E, McKay D, Balakrishnan R, Radjai M, DiPasquantonio V, Shoda D, Hatch R, Venkatasubramanian K (1990) Genetic engineering of metabolic pathways applied to the production of phenylalanine. Ann NY Acad Sci 589:16–24
126. Rüffer N, Heidersdorf U, Kretzers I, Sprenger GA, Raeven L, Takors R (2004) Fully integrated L-phenylalanine separation and concentration using reactive-extraction with liquid-liquid centrifuges in a fed-batch process with *E. coli*. Bioprocess Biosyst Eng 26:239–248
127. Konstantinov KB, Nishio N, Seki T, Yoshida T (1991) Physiologically motivated strategies for control of the fed-batch cultivation of recombinant *Escherichia coli* for phenylalanine production. J Ferment Bioeng 71:350–355
128. Liu SP, Xiao MR, Zhang L, Xu J, Ding ZY, Gu ZH, Shi GY (2013) Production of L-phenylalanine from glucose by metabolic engineering of wild type *Escherichia coli* W3110. Proc Biochem 48:413–419
129. Zhou H, Liao X, Wang T, Du G, Chen J (2010) Enhanced L-phenylalanine biosynthesis by co-expression of pheAfbr and aroFwt. Bioresour Technol 101:4151–4156
130. Dobson R, Gray V, Rumbold K (2012) Microbial utilization of crude glycerol for the production of value-added products. J Ind Microbiol Biotechnol 39(2):217–226
131. Gottlieb K, Albermann C, Sprenger GA (2014) Improvement of L-phenylalanine production from glycerol by recombinant *Escherichia coli* strains: the role of extra copies of *glpK*, *glpX* and *tktA* genes. Microb Cell Fact 13(1):96
132. Thongchuang M, Pongsawasdi P, Chisti Y, Packdibamrung K (2012) Design of a recombinant *Escherichia Coli* for producing L-phenylalanine from glycerol. World J Microbiol Biotechnol 28(10):2937–43
133. Weiner M, Tröndle J, Albermann C, Sprenger GA, Weuster-Botz D (2014) Carbon storage in recombinant *Escherichia Coli* during growth on glycerol and lactic acid. Biotechnol Bioeng 111(12):2508–2519

134. Weiner M, Albermann C, Gottlieb K, Sprenger GA, Weuster-Botz D (2014) Fed-batch production of L-phenylalanine from glycerol and ammonia with recombinant *Escherichia coli*. Biochem Eng J 83:62–69
135. Gerigk M, Bujnicki R, Ganpo-Nkwenkwa E, Bongaerts J, Sprenger G, Takors R (2002) Process control for enhanced L-phenylalanine production using different recombinant *Escherichia coli* strains. Biotechnol Bioeng 80(7):746–754
136. Jenzsch M, Gnoth S, Beck M, Kleinschmidt M, Simutis R, Lübbert A (2006) Open-loop control of the biomass concentration within the growth phase of recombinant protein production processes. J Biotech 127:84–94
137. Kikuchi Y, Tsujimoto K, Kurahashi O (1997) Mutational analysis of the feedback sites of phenylalanine-sensitive 3-deoxy-D-arabino-heptulosonate-7-phosphate synthase of *Escherichia coli*. Appl Environ Microbiol 63(2):761–762
138. Zhang X, Shanmugam KT, Ingram LO (2010) Fermentation of glycerol to succinate by metabolically engineered strains of *Escherichia coli*. Appl Environ Microbiol 76:2397–2401
139. Schürmann M, Sprenger GA (2001) Fructose-6-phosphate aldolase is a novel class I aldolase from *Escherichia coli* and is related to a novel group of bacterial transaldolases. J Biol Chem 276:11055–11061
140. Xu YF, Amador-Noguez D, Reaves ML, Feng XJ, Rabinowitz JD (2012) Ultrasensitive regulation of anapleurosis via allosteric activation of PEP carboxylase. Nat Chem Biol 22, 8 (6):562–568.
141. Kotlarz D, Buc H (1982) Phosphofructokinases from *Escherichia coli*. Methods Enzymol 90:60–70

Adv Biochem Eng Biotechnol (2016) 152: 137–151
DOI 10.1007/10_2014_293

Published Online: 31 January 2015

Advances and Practices of Bioprocess Scale-up

Jianye Xia, Guan Wang, Jihan Lin, Yonghong Wang, Ju Chu, Yingping Zhuang and Siliang Zhang

Abstract This chapter addresses the update progress in bioprocess engineering. In addition to an overview of the theory of multi-scale analysis for fermentation process, examples of scale-up practice combining microbial physiological parameters with bioreactor fluid dynamics are also described. Furthermore, the methodology for process optimization and bioreactor scale-up by integrating fluid dynamics with biokinetics is highlighted. In addition to a short review of the heterogeneous environment in large-scale bioreactor and its effect, a scale-down strategy for investigating this issue is addressed. Mathematical models and simulation methodology for integrating flow field in the reactor and microbial kinetics response are described. Finally, a comprehensive discussion on the advantages and challenges of the model-driven scale-up method is given at the end of this chapter.

Keywords Bioprocess engineering · Scale-up · Bioprocess monitoring · Fluid dynamics · Microbial kinetics · Mathematical model

Contents

1 Introduction 138
2 Bioprocess Research Based on Multi-scale Analysis of Fermentation Process 139
2.1 Theory of Multi-scale Analysis for Fermentation Process 139
2.2 Process Optimization in the Context of Association Analysis of Process Parameters 141
2.3 Successful Scale-up by Combination of Process Parameters and Fluid Dynamics 142
3 Optimization and Scale-up of Bioprocess Based on Integration of Fluid Dynamics and Biokinetics 143
3.1 Heterogeneity Environment in Large-scale Bioreactor and Its Impact on Cells 143
3.2 Development of Scale-down System and Its Applications 144
3.3 Model-driven Rational Scale-up of Bioprocess 145
4 Conclusions 147
References 148

J. Xia · G. Wang · J. Lin · Y. Wang · J. Chu · Y. Zhuang · S. Zhang (✉)
State Key Laboratory of Bioreactor Engineering,
East China University of Science and Technology,
130 Meilong Road, Shanghai 200237, China
e-mail: siliangz@ecust.edu.cn

1 Introduction

With the ever-growing demand for foods, nutriceuticals, pharmaceuticals, fuels, and materials as well as for sustainable development of economy and environment, microbial fermentations using low-cost and renewable feedstocks have become increasingly important. Great effort has been endeavored for improving yields, titers, and productivities of aimed products through bioprocess engineering strategies since the very early beginning of the fermentation industry. Among them, optimization and scale-up method toward industrial process shows great importance. In principle, fermentation capacity is of great extent depending on two compelling parts, i.e., the cellular machinery (determined by gene functions and individual enzyme kinetics) and the extracellular environment (determined by fluid dynamics in bioreactor) [1]. In this regard, it is a permanent task and challenge for biotechnological research to discover and gain more knowledge of cell kinetics and bioreactor fluid dynamics, and interaction between these two parts, in order to accelerate the transition process from laboratory investigation to industrial application [2].

In the past four decades, a huge amount of advanced process monitoring techniques have been developed and applied for bioprocess monitoring and control, e.g., IR spectroscopy for online real-time measurement of glucose, glutamate, fructose, glutamine, proline, and ammonia [3]; Raman spectroscopy for measurement of glucose, acetate, formate, lactate and phenylalanine, and carotenoid production [4]; capacitance sensor for biomass measurement [5]; and online MS for measurement of real-time concentration of O_2 and CO_2 in exhaust gas [6]. These rich real-time data, together with the increasing mega data along with the advent of OMICS techniques, make tens of thousands of parameters can be monitored and analyzed simultaneously [7, 8] at different scales. However, if a certain phenomenon happening on one scale is regarded as the main research objective, it is hard to discover the relevance among different scales because of difficulties with micro- and macro-statistical processing of the data. Thus the multi-scale analysis methodology on bioprocess becomes an issue of interest [9].

Another important issue of bioprocess scale-up is heterogeneous flow field in the industrial-scale bioreactor and its negative effect on the cell physiology [10, 11]. Bioprocesses conducted in large-scale bioreactors always face mixing or mass transfer problem which may not occur in laboratory- or bench-scale bioreactors. It has been observed that biomass yield of *E. coli* decreased by 20 % in large-scale bioreactor (12 m^3) compared to in bench-scale bioreactor [12]. Heterogeneous environment in the large-scale bioreactor is deemed to main cause of the scale-up issue; thus, experimental investigation using scale-down system in laboratory scale was proposed to gain more insights on this issue. Neubauer [13] reviewed different scale-down simulation systems and concluded that the scale-down system provided improved possibilities to evaluate how a bioprocess would behave in the final industrial scale. However, lack of quantitative information of flow field both in large-scale bioreactor and scale-down system makes it hard to determine whether they are under the same heterogeneity. Computational fluid dynamics (CFD) tools can possibly be the best solution to this problem, but few researches have been reported regarding this. Using the multi-scale analysis technique, more information about the cell physiology under different scales can be gained. Combined with flow field information in different scale bioreactors by CFD, it can be used to rationally direct the bioprocess scale-up [14]. To some extent, this can solve the real industrial problem quickly without deep insight into the cell kinetic mechanism.

Mathematical models have been used to understand, predict, and optimize the properties and behavior of cells and the bioprocess [15]. Kinetic models can be used for different purposes, e.g., enhancing substrate utilization and product yield, detecting metabolic engineering target, and improving process design [16]. For rational scale-up of bioprocess, kinetic models that predict cell behavior in dynamic external environments coupled with fluid dynamics models that describe mass transfer and mixing in the bioreactor can be employed to predict output of the whole culture system under different scales. This issue was raised by Vrabel et al. [17] and Schamalzriedt et al. In their research using Euler–Euler simulation framework, however, historical effects of external environment on cells were not considered. Lapin [18, 19] solved this problem by using Euler–Lagrange simulation framework, which simulated the biophase using Lagrangian frame of reference. However, much more precise and fast dynamic response experiments on cells should be designed to validate the proposed models.

In this chapter, we discuss mainly the advances and applications of bioprocess techniques in the bioindustrial field. New improvement of the multi-scale analysis of fermentation process and its successful practices in industrial bioprocess were addressed. Cause of heterogeneity in large-scale bioreactor and its effect on microbial physiology will be analyzed first, and then, experimental scale-down investigation method and mathematical modeling are reviewed to decipher this issue. The development and new perspective of bioprocess engineering are also proposed at the end of this chapter.

2 Bioprocess Research Based on Multi-scale Analysis of Fermentation Process

2.1 Theory of Multi-scale Analysis for Fermentation Process

Nowadays, to our knowledge, strain improvement strategy relies stepwise on an oriented genome-scale restructure, rather than random mutagenesis. Following the acquired high-yield strain, how to effectively and efficiently apply it and scale-up to industrial-scale challenges both academic and industrial researchers. Traditional scale-up is mainly based on the principle of similarity and dimensional analysis, such as using the same specific power consumption, K_La, impeller speed, and mixing time as scale-up criteria. Under this circumstance, however, no biological properties are taken into account, and thus, no universal principle can be deduced for successful scale-up [20]. In other words, biologically fermentation system is a complex system with a large number of elements, building blocks, or agents, capable of buffering stimuli with one another and with their environments [21]. To address this problem, in the wake of a long-term accumulation of theory and practice, the multi-scale analysis method dealing with the complex interplay between the cellular machinery and the extracellular environment was at the first time proposed by Zhang et al. [22] and has already successfully found its applicability in the production of varying commercial products, such as penicillin, erythromycin, chlortetracycline, inosine, guanosine, recombinant human serum, and malaria vaccine.

It should be noted that output of real fermentation process is limited by strain capacity but driven by nutrients environment that surrounding microbial cells, especially in industrial-scale bioreactor. For instance, it is necessary to maintain the dissolved oxygen (DO) tension above 30 % during penicillin fermentation; otherwise, the penicillin synthesis is irreversibly affected [23]. While, in other cases, oxygen limitation strategy is very helpful, e.g., during glucoamylase fermentation using *Aspergillus niger* [24]. Most interesting is the fact that there are some robust strains and wild-type or engineered strains, less sensitive to environmental changes in a certain range, which to a great extent relieves the burden of process optimization and scale-up [25, 26]. In most cases, however, microbial physiology and product formation are prone to be influenced by changes in the extracellular environment. Nowadays, there is an increasing tendency of studying effects of environmental gradients on metabolism, flux, and growth rate [25, 27–29]. Due to insufficient mixing and mass transfer limitations in large-scale fermentors, various gradients, including substrate gradient, DO gradient, pH gradient, and carbon dioxide gradient, do occur inevitably and the cells are always experiencing feast/famine cycles. As a consequence, biomass, productivity, and yield are severely affected [30]. As first proposed by Oosterhuis, the scale-down simulation experiment has been employed to investigate local limitation effects in the large-scale fermentors, accordingly. Since then, a couple of scale-down simulation systems have been designed, which could accelerate the prediction of large-scale performance [13, 20, 31, 32]. It is, therefore, of great significance to detect environmental changes in the course of fermentation and adjust corresponding regimes, such as feeding strategy, to enhance productivity and yield, afterward [33].

Aided by numerous instruments mounted on the fermentor, the multi-scale method using many online critical parameters, such as DO, pH, temperature (T), oxygen uptake rate (OUR), carbon dioxide emission rate (CER), and respiratory quotient (RQ), depicts a holistic picture of fermentation process. In other words, dynamic tendency of bioprocess is very important for getting a deeper insight into microscopic metabolism of the culture over the entire process. For example, biomass is one of the most important physiological parameters in fermentation process, which is usually expressed as dry cell weight (DCW). However, DCW, the aggregation of both active and inactive cells and hence, fails to reflect real-time microbial physiology. Colony-forming units (CFU) is thus applied to estimate the viable cells, but it is time consuming and difficult to know what happens in a fermentor as a function of time. Thus, online capacitance probe was used to get accurate determination of viable cells and reflects time-course of active biomass, effectively and efficiently [5]. With increasing amount of biosensors being available for detecting changes in a real bioprocess, the multi-scale analysis method using the correlation and tendency of process parameters will be a more powerful methodology in studying practices and developing optimized strategies.

2.2 Process Optimization in the Context of Association Analysis of Process Parameters

The multi-scale analysis method describing the highly correlated process parameters has been employed to analyze dynamics of the culture. On one hand, biosensors mounted in the fermentor can read direct parameters such as DO, T, stirrer speed, and gas flow rate; on the other hand, indirect parameters standing for time-course physiology and metabolism of the strain such as OUR, CER, and RQ can be on line calculated with the data acquisition software package. A number of successful cases relating to bioprocess scale-up were realized by the multi-scale analysis method.

As a case in point, in the late-phase production of guanosine, a typical phenomenon was that OUR, CER, productivity, and yield of guanosine were rapidly decreasing, while sugar and ammonia consumption were still increasing. Considering the carbon balance, it was deduced that some intermediates of metabolism, such as amino acids, organic acids, or some other nitrogenous substances, had accumulated, and the metabolic flux had shifted to synthesis of these intermediates, other than guanosine. As evidenced by the time-course of by-products, key enzymatic activities, and stoichiometric calculation of metabolic flux shift, it was concluded that late-phase production of guanosine competed against alanine accumulation, which provided possible direction for metabolic engineering and also for process optimization [6, 34, 35]. Another successful case was the optimization of recombinant human serum albumin by an engineered *P. pastoris*. Generally, a severe phenomenon was the emergence of zero dissolved oxygen level after fermentation for 36 h, which was the main problem to be solved. A novel medium and feeding strategy were proposed to address this problem by analysis of online parameters using the multi-scale analysis method [9].

In parallel, the multi-scale analysis method was also successfully applied to optimization of the production of secondary metabolites, such as erythromycin, penicillin, avermectin, and tylosin. Penicillin is the first bio-based pharmaceutical discovered by Alexander Fleming, and its production titer nowadays reaches over 100,000 times higher than the original strain [36]. However, from stoichiometric perspective, the theoretical value was calculated to be 0.50 mol/mol glucose, where of course, direct sulfhydrylation was only considered and no by-product was taken into account [37]. Therefore, it is still a long way to optimize the strain and process strategy. For example, on a large-scale penicillin production, one of the most important regulation strategies is breakup of foam by frequent addition of antifoam reagent, soybean oil. Soybean oil in fact has other mechanisms of function because its addition brings either positive or negative effects if not properly designed. To this end, the multi-scale analysis method was used to analyze the underlying mechanism of addition of soybean oil. In conclusion, soybean oil may be: (a) used as carbon source. One mole of oil is equivalent to more than two moles of glucose; hence, the addition of a large amount of oil will result in abundant carbon source, which in turn influence the specific growth rate and thus impair final productivity. (b) correlated with oxygen supply. An appropriate amount of foam is in favor of oxygen transfer especially in early stage of fermentation. Moreover, another effect of soybean oil drives carbon dioxide out of foam and reduces its side effect. As a consequence, it is suggested that soybean oil should be intermittently added in small amounts.

Derived from the multi-scale analysis method, "chaotic phenomenon" was frequently used for characterization of secondary metabolism. What is the "chaotic phenomenon"? After years' study of synthesis of secondary metabolites, e.g., antibiotics, a general phenomenon is frequently happening in both laboratory and industry cases that the initial fermentation phase shows process polymorphism and uncertainty, and tiny difference of initial input may exert a huge impact on the final result. For instance, the "chaotic phenomenon" did occur in the production of erythromycin when different carbon sources were used. It was drawn from the data-trend curves that soybean meal, instead of glucose, via amino acid metabolism might function in supplying erythromycin with the carbon skeleton. Properly controlled, the productivity and yield of erythromycin would be noticeably enhanced [38]. In parallel with discovery of the "chaotic phenomenon," stringent response, a common metabolic regulation mechanism found in a wide range of prokaryotes and also in plants is involved in the synthesis and accumulation of guanosine tetra-(ppGpp) or pentaphosphates (pppGpp) when cells encounter nutrient starvation. This phenomenon, a global transcriptional regulation response, has considerable functions in growth rate control, DNA maintenance, protein turnover, sporulation, and also in antibiotic synthesis [39–42]. Therefore, the multi-scale analysis method accelerates profound comprehension of the bioprocess and may also provide clues for explanation of some underlying mechanisms involving the molecular scale (genetics), the cellular scale (metabolic regulation), and the reactor scale (process control). It can be seen that the multi-scale analysis method provides series of principles for investigating the fermentation process, which can be seen as a complex system [43].

2.3 Successful Scale-up by Combination of Process Parameters and Fluid Dynamics

Two important issues, namely physiology of microbes and flow field in fermentors, are highly interconnected during the whole fermentation process and affect final fermentation performance. Of great importance is to understand relationship between these two aspects, which will further accelerate the scale-up process [44]. Following the multi-scale analysis method describing real-time physiology of cells, a huge amount of efforts have been taken to understand flow field of fermentation system. CFD has been established as a very useful tool in solving serial problems such as flow of fluid, mixing, transfer, and chemical reactions [45, 46]. CFD technology follows the fundamentals where the conservation of fluid mass, momentum, and energy are governed, and it has been used to find out the bottleneck involved in industrial scale-up.

As a case in point, on a 12 m^3 cephalosporin C (CPC) production scale, Yang et al. simulated two different impeller combinations using CFD, in which one was the conventional radial impellers and the other was the combination of radial and axial impellers. From the point of simulation results, it was conceivable that the novel impeller configuration generated more homogeneous conditions. In industrial practice, soybean oil is widely used as carbon source in the antibiotics fermentation process. Though it is poorly soluble in broth, soybean oil is normally fed at the top of fermentor, which results in oil gradient in case of poor mixing. The discrepancy of process parameter RQ value under these two impeller configurations reveals different consumption rate of soybean oil. It was found that RQ profiles branched after the first 50 h fermentation between the two impeller combinations, and RQ value under novel impeller combination was much closer to the theoretical value. This indicates that more homogeneous soybean oil concentration was formed under novel impeller combination due to its more effective mixing capacity. Moreover, hydrodynamic environment generated by the novel combination may favor formation of dispersed arthrospores, rather than mycelia, and thus enhance CPC production [47, 48].

In 132 m^3 erythromycin fermentation, by using CFD simulation of fluid dynamics on different scales, it was concluded that the main cause of impairment of physiological metabolism and erythromycin formation was the decrease of OTR as volume increased. The OUR in 132-m^3 fermentor was obviously lower than that in 50-L fermentor, which further confirmed the insufficient oxygen supply in the larger-scale bioreactor. Therefore, CFD helps to understand the relationship between flow field in fermentors and physiology of microbes [14]. In conclusion, integration of fluid dynamics and process parameters has been proven to be an efficacious way to comprehensively understand what is going on in different scale fermentors [44].

3 Optimization and Scale-up of Bioprocess Based on Integration of Fluid Dynamics and Biokinetics

3.1 Heterogeneity Environment in Large-scale Bioreactor and Its Impact on Cells

A dynamic environment with large spatial and temporal heterogeneities is always produced in large-scale fermentor especially under fed-batch mode [49]. Several aspects will change inevitably during the scale-up from bench- to industrial-scale [31].

For example, with increasing liquid height, hydrostatic pressure at bottom of the large reactor can easily reach over 1 bar, which in turn results in higher oxygen solubility, and vertical oxygen concentration gradients will then be formed. Mixing time can also easily reach over 2 min in large-scale fermentor up to 19 m^3 [50], while it is only in the order of seconds in laboratory-scale bioreactor. Furthermore, pH gradient and temperature gradient are also found in large-scale bioreactors. As Lara et al. [10] has analyzed, when mixing time in the large scale was in the same order of magnitude as the reaction time, heterogeneities can be anticipated. However, it is hard to maintain constant mixing time at different scales because of limitation of power input in the large-scale bioreactor.

It should be noted that cells are alive and highly responsive compared to chemical reactant or catalyst. These above-mentioned heterogeneities can induce multiple physiological responses; different cells have different responses according to their genetic or metabolic control mechanism. As bioreactor scale increases, mixing time may exceed over the bioreaction characteristic time, which, however, is independent of bioreactor size [10, 49]. Thus, cells in large-scale bioreactors may probably experience an oscillating environment which leads to a heterogenetic cell population. This usually causes lower yields and productivities and an increased by-product formation compared to laboratory-scale bioreactors [11, 12, 29]. Therefore, lots of time and effort are taken to study scale-up process, which is a great hurdle for rapid development of bioprocesses from micro-liter cultures to the industrial scale [51]. Understanding the heterogeneity in industrial-scale bioreactors and the performance of cells exposed to an alternating environment is of great importance for efficient enhancement of biotechnical processes.

3.2 *Development of Scale-down System and Its Applications*

The scale-down concept was first proposed by Kossens and coworkers [52]. Regime analysis was used to interpret the large-scale process by comparing characteristic time of different processes, and scale-down simulation system was then established based on this information [53]. It is proven to be an effective way for investigating the effect of oscillating environments on cell metabolic and physiological behavior in a laboratory-scale equipment. Various scale-down systems have been designed to investigate effects of heterogeneities such as pH, temperature, dissolved oxygen, and substrate on physiology of bacteria [54], yeast [55], filamentous fungi [56], and mammalian cells [57].

Among all kinds of scale-down systems mimicking environmental gradients, the most frequently used can be roughly classified into two groups, namely one-compartment systems and two-compartment systems. Usually, a one-compartment system consists of a stirred tank reactor (STR) [58] or specially designed tubular reactor [59]. A one-compartment scale-down system was designed to simulate dissolved oxygen tension gradient by León-Rodríguez et al. [60]. The results showed that one-compartment system was able to generate sustained oscillating dissolved oxygen profiles. One-compartment system is simple in structure and relies on control strategy to generate gradients. However, in one-compartment scale-down system, all cells are in the same condition which means such approach is unable to mimic the cell population heterogeneity in real bioreactor. While, in a two-compartment system, cells circulate through the two connected compartments, each of which implements a certain different condition. The two-compartment scale-down system can be constructed by either connecting two STRs or connecting one STR with a plug-flow reactor (PFR), which can be used to simulate real cycling times and environmental gradients. The combination of two compartments makes the system more facile to reflect the real large-scale bioreactor. Junne et al. [61] constructed a two-compartment bioreactor with commercially available parts to investigate a non-sporulating *B. subtilis* response to oscillating glucose and dissolved oxygen concentration and found that carbon flux at excess glucose and low DO concentration was shifted toward ethanol formation, as a result, diminished glucose uptake and altered amino acid synthesis were observed. Sandoval-Basurto et al. [62] implemented a two compartment scale-down system to investigate dissolved oxygen gradient effect on recombinant *E. coli*. Two STRs with different DO levels were connected. It was found that acetic, lactic, formic, and succinic acids accumulated under oscillating DO conditions.

For scale-down system, it is very important to make sure that it can represent the situation that cells encounter in large-scale bioreactor. Otherwise, the investigation of different oscillating conditions may not be relevant to the key hurdle of the scale-up. To this end, CFD investigation on the large-scale reactor can pave the way for rational design of the scale-down system, as it can provide more detailed and local information about hydrodynamics and mixing. However, the real fluctuating environment in the large-scale bioreactor is a result of both fluid mixing and metabolic response of cells in the reactor. This means not only the fluid dynamics in the reactor, but also kinetic information of the cells should be considered for this purpose. Therefore, an idea of coupling these two aspects comes out for better understanding of the scale-up process.

3.3 Model-driven Rational Scale-up of Bioprocess

Mathematical models describing both fluid dynamics in the reactor and bioreaction kinetics will help to shed bright light on the behavior of the industrial-scale fermentation system, which in the end leads to a rational design of an industrial fermentation process. Since 1980s, this method has been used to simulate the performance of large-scale fermentation system by coupling fluid dynamics model with simple substrate uptake model [63]. However, the complexity of both models makes it more difficult to be conducted in engineering practice; hence, simplified fluid dynamics model of compartment model approach and Monod-type kinetic model were always used at that time [17]. With rapid development of computer hardware and simulation algorithm, nowadays, the CFD software has been used to solve flow field problems up to 120-m^3 scale reactor [14]. Integrated models that couple fluid dynamics and microbial kinetics have been used to simulate hydrodynamic effects on filamentous morphology [64] and physiological response [65] in different kinds of reactors.

All the above-mentioned simulation work are based on Euler–Euler approach in which gas, liquid, and biophase are all considered as a continuum and described in terms of their volume fractions. In the past two decades, this approach has been applied to various biological processes [1, 30, 66]. A simple multi-scale kinetic model based on Herbert's concept was coupled with a CFD model to investigate influence of mixing mechanism in a 1.5-L bioreactor on ethanol fermentation. Compared to experiment results, simulated data showed approximately 5 % error for yield and 14 % error for productivity [67]. Elqotbi successfully simulated the whole 60 h fermentation process of *A. niger* with limited computational efforts based on the separated solution of the flow field, the mass transfer, and microbial reaction. Such stepwise solution strategy divides the whole fermentation process into three steps—firstly solving the fluid flow field, secondly imposing oxygen mass transfer and bioreaction, and finally updating flow field after bioreaction. This strategy successfully solved the problem of different timescales between hydrodynamics and bioreaction [68]. Segregated model for the biophase using population balance model (PBM) is also coupled to CFD model to address bioreactor-scale effect on the cell population heterogeneity in a recent published work by Morchain et al. [69], which is also based on Euler–Euler frame method.

In fact, the Euler–Euler approach leads to loss of realism if individual history of cells becomes the focus of attention, e.g., when considering cumulative starvation effects in cells during fed-batch fermentation [18]. In contrast, the Euler–Lagrange approach overcomes this problem, in which the fluid phase is treated as a continuum whereas the dispersed biophase is solved by tracking a large number of particles through the calculated flow field [70]. By using this approach, the analysis of lifeline of individual cell in space and time is possible. Lapin et al. [18] first employed this approach to couple yeast glycolytic oscillation with turbulent flow in a 68-L reactor to demonstrate the influence of bioreactor mixing intensity to the synchronization of yeast glycolytic oscillation in a population level. Their results showed that non-ideal mixing condition (N_{imp} = 55 rpm) resulted in slightly diminished degree of synchrony as compared to ideally mixing case (N_{imp} = 165 rpm). They used this approach again to simulate fed-batch fermentation of *E. coli* in a 900-L bioreactor [19]. In this model, a sugar uptake kinetic model (phosphotransferase system, PTS) was coupled to turbulent flow in bioreactor. The activity of the sugar uptake system depends on the local concentration of glucose as well as the ratio of the intracellular concentrations of phosphoenolpyruvate and pyruvate, which in turn is a function of the history of individual cell. Heterogeneity on the specific sugar uptake rate was observed among the *E. coli* population. Much of their results were only indirectly verified by experimental observations, but the proposed simulation framework in their work was to some extent a promising method for better understanding the scale-up problem.

To be applied for rational design and scale-up of bioprocess, the integration simulation approach coupling fluid dynamics in bioreactor and cellular kinetics based on both the Euler–Euler and the Euler–Lagrange shows attractive potential. But, there is still a long way to go. One main challenge of this method is how to keep balance of computational expense and simulation accuracy. Another problem of this method is validation of simulation results, even though there are some reports on measurement of intracellular metabolites in single cell level [71], it is really a hard work to get sufficient validation data in even laboratory-scale bioreactor at the present state.

4 Conclusions

Despite the central role of scale-up issue in biotechnology and the large body of literatures, there seems to be no common, universally applicable strategy [72]. It has been ever stated that scale-up is an art rather than an exact science [73]. Indeed, the fermentation process conducted in bioreactor is really a complex system, as the cell, which is alive, has a precise control mechanism which shows different responses to environment perturbations on different scales. That is believed to be the main cause of various scale-up problems.

With the great increase of knowledge of the interplay between cell physiological response and extracellular nutrient conditions, we are approaching more rational scaling up of the bioprocess. To identify process-specific stress factors and to understand the physiological responses to the vessel specific physical conditions, the mutual influences and interactions of the various physical and physiological parameters need to be analyzed in detail [72]. Multi-scale fermentation analytical method coupled with fluid dynamics investigation can promisingly implement this goal and has been applied to optimize and scale-up of different fermentation process, which was proved to be an efficient approach. A holistic scale-up strategy consists of a comprehensive and detailed process characterization of both laboratory-scale and industrial-scale fermentor to identify key process parameters influencing product yield and productivity.

In parallel, with development of both metabolic engineering and systems biotechnology, more and more mathematical models describing cell metabolism and its regulation mechanism have been proposed. But most of them focus on the stoichiometry relations but not kinetic effects. As Almquist et al. [16] pointed out, kinetic model should be a powerful tool for better understanding the metabolic mechanism for their response to either genetic manipulation or environment fluctuations. While it is a challenge to build a holistic realistic kinetic model as little is known about the in vivo mechanisms of enzymes and transporters. Short-term perturbation experiment based on fast sampling technique and precise measurement of intracellular metabolites is a well-performed tool for establishing such in vivo kinetic model. Such model with careful validation can be used to couple CFD model in different scale reactors for rational scale-up of different bioprocesses.

Acknowledgment Financial support of State Key Development Program of Basic Research of China (973 Program, Grant No. 2013CB733600), NWO-MoST Joint Program (2013DFG32630), National High Technology Research and Development Program of China (863 Program, Grant No. 2012AA021201 and 2014AA021701), and China Ministry of Science and Technology under Contract (Grant No. 2012BAI44G00) are gratefully acknowledged.

References

1. Schamalzriedt S et al (2003) Integration of physiology and fluid dynamics. In: Scheper T (ed) Advances in biochemical engineering/biotechnology. Springer, Berlin, pp 19–68
2. Babel W, Brinkmann U, Müller R (1993) The auxiliary substrate concept—an approach for overcoming limits of microbial performances. Acta Biotechnol 13(3):211–242
3. Landgrebe D et al (2010) On-line infrared spectroscopy for bioprocess monitoring. Appl Microbiol Biotechnol 88(1):11–22
4. Lee HLT et al (2004) In situ bioprocess monitoring of Escherichia coli bioreactions using Raman spectroscopy. Vibrational Spectroscopy 35(1–2):131–137
5. Li L et al (2014) Optimization of polyhydroxyalkanoates fermentations with on-line capacitance measurement. Bioresour Technol 156:216–221
6. Zhang SL et al (2004) Studies on the multi-scale problems of guanosine fermentation process. Chin J Bioprocess Eng 2(3):23–29
7. Bonneau R et al (2007) A predictive model for transcriptional control of physiology in a free living cell. Cell 131(7):1354–1356
8. Ishii N et al (2007) Multiple high-throughput analyses monitor the response of *E. coli* to perturbations. Science 316:593–597
9. Zhang SL, Chu J, Zhuang Y (2004) A multi-scale study of industrial fermentation processes and their optimization. Adv Biochem Eng Biotechnol 87:97–150
10. Lara AR et al (2006) Living with heterogeneities in bioreactors. Mol Biotechnol 34 (3):355–381
11. Enfors S-O et al (2001) Physiological responses to mixing in large scale bioreactors. J Biotechnol 85(2):175–185
12. Bylund F et al (1998) Substrate gradient formation in the large-scale bioreactor lowers cell yield and increases by-product formation. Bioprocess Eng 18(3):171–180
13. Neubauer P, Junne S (2010) Scale-down simulators for metabolic analysis of large-scale bioprocesses. Curr Opin Biotechnol 21(1):114–121
14. Zou X et al (2012) Real-time fluid dynamics investigation and physiological response for erythromycin fermentation scale-up from 50 L to 132 m^3 fermenter. Bioprocess Biosyst Eng 35(5):789–800
15. Tyo KEJ, Kocharin K, Nielsen J (2010) Toward design-based engineering of industrial microbes. Curr Opin Microbiol 13(3):255–262
16. Almquist J et al (2014) Kinetic models in industrial biotechnology—improving cell factory performance. Metab Eng 24:38–60
17. Vrabel P et al (2001) CMA: integration of fluid dynamics and microbial kinetics in modelling of large-scale fermentations. Chem Eng J 84(3):463–474
18. Lapin A, Müller D, Reuss M (2004) Dynamic behavior of microbial populations in stirred bioreactors simulated with euler–lagrange methods: traveling along the lifelines of single cells. Ind Eng Chem Res 43(16):4647–4656
19. Lapin A, Schmid J, Reuss M (2006) Modeling the dynamics of E. coli populations in the three-dimensional turbulent field of a stirred-tank bioreactor–A structured-segregated approach. Chem Eng Sci 61(14):4783–4797
20. Wang G et al (2014) Prelude to rational scale-up of penicillin production: a scale-down study. Appl Microbiol Biotechnol 98(6):2359–2369

21. Ottino JM (2003) Complex systems. AIChE J 49(2):292–299
22. Zhang SL, Chu J, Zhuang YP (2004) A multi-scale study of industrial fermentation processes and their optimization. In: Biomanufacturing. Springer, Berlin, pp 97–150
23. Vardar F, Lilly M (1982) Effect of cycling dissolved oxygen concentrations on product formation in penicillin fermentations. Eur J Appl Microbiol Biotechnol 14(4):203–211
24. Pedersen L et al (2012) Industrial glucoamylase fed-batch benefits from oxygen limitation and high osmolarity. Biotechnol Bioeng 109(1):116–124
25. Baez A et al (2011) Simulation of dissolved CO_2 gradients in a scale-down system: a metabolic and transcriptional study of recombinant Escherichia coli. Biotechnol J 6 (8):959–967
26. Käß F et al (2013) Assessment of robustness against dissolved oxygen/substrate oscillations for C. glutamicum DM1933 in two-compartment bioreactor. Bioprocess Biosyst Eng 1–12
27. Taymaz-Nikerel H et al (2013) Changes in substrate availability in Escherichia coli lead to rapid metabolite, flux and growth rate responses. Metabol Eng 16:115–129
28. de Jonge LP et al (2011) Scale-down of penicillin production in Penicillium chrysogenum. Biotechnol J 6(8):944–958
29. Käß F et al (2014) Process inhomogeneity leads to rapid side product turnover in cultivation of Corynebacterium glutamicum. Microb Cell Fact 13(1):1–11
30. Larsson G et al (1996) Substrate gradients in bioreactors: origin and consequences. Bioprocess Eng 14(6):281–289
31. Takors R (2012) Scale-up of microbial processes: impacts, tools and open questions. J Biotechnol 160(1–2):3–9
32. Noorman H (2011) An industrial perspective on bioreactor scale-down: what we can learn from combined large-scale bioprocess and model fluid studies. Biotechnol J 6(8):934–943
33. de Jonge L et al (2014) Flux response of glycolysis and storage metabolism during rapid feast/famine conditions in Penicillium chrysogenum using dynamic ^{13}C labeling. Biotechnol J 9 (3):372–385
34. Chu J et al (2005) Correlation between key enzyme activities in the inosine synthetic pathway and inosine production. Process Biochem 40(2):891–894
35. Chen SX et al (2005) Enhancement of inosine production by *Bacillus subtilis* through suppression of carbon overflow by sodium citrate. Biotechnol Lett 27(10):689–692
36. Rokem JS, Lantz AE, Nielsen J (2007) Systems biology of antibiotic production by microorganisms. Nat Prod Rep 24(6):1262–1287
37. Jorgensen H et al (1995) Metabolic flux distributions in Penicillium chrysogenum during fed-batch cultivations. Biotechnol Bioeng 46(2):117–131
38. Zhao J et al (2002) Study on chaos of the erythromycin fermentation process. Chin J Antibiot 27(12):717–719
39. Chatterji D, Kumar Ojha A (2001) Revisiting the stringent response, ppGpp and starvation signaling. Curr Opin Microbiol 4(2):160–165
40. Kang SG et al (1998) Actinorhodin and undecylprodigiosin production in wild-type and relA mutant strains of Streptomyces coelicolor A3 (2) grown in continuous culture. FEMS Microbiol Lett 168(2):221–226
41. Hoyt S, Jones GH (1999) relA is required for actinomycin production in Streptomyces antibioticus. J Bacteriol 181(12):3824–3829
42. Kuhar I, Žgur-Bertok D (1999) Transcription Regulation of the Colicin Kcka Gene Reveals Induction of Colicin Synthesis by Differential Responses to Environmental Signals. J Bacteriol 181(23):7373–7380
43. Ye X et al (2005) Multi-scale methodology: a key to deciphering systems biology. Front Biosci 10:961–965
44. Wang Y et al (2009) Industrial bioprocess control and optimization in the context of systems biotechnology. Biotechnol Adv 27(6):989–995
45. Sharma C, Malhotra D, Rathore AS (2011) Review of computational fluid dynamics applications in biotechnology processes. Biotechnol Prog 27(6):1497–1510

46. Xia JY et al (2008) Computational investigation of fluid dynamics in a recently developed centrifugal impeller bioreactor. Biochem Eng J 38(3):406–413
47. Xia JY et al (2009) Fluid dynamics investigation of variant impeller combinations by simulation and fermentation experiment. Biochem Eng J 43(3):252–260
48. Yang YM et al (2012) A novel impeller configuration to improve fungal physiology performance and energy conservation for cephalosporin C production. J Biotechnol 161 (3):250–256
49. Hewitt CJ, Nienow AW (2007) The scale up of microbial batch and fed-batch fermentation processes. Adv Appl Microbiol 62:105–135
50. Junker BH (2004) Scale-up methodologies for Escherichia coli and yeast fermentation processes. J Biosci Bioeng 97(6):347–364
51. Neubauer P et al (2013) Consistent development of bioprocesses from microliter cultures to the industrial scale. Engineering in Life Sciences 13(3):224–238
52. Oosterhuis NMG et al (1985) Scale-down and optimization studies of the gluconic acid fermentation by Gluconobacter oxydans. Biotechnol Bioeng 27(5):711–720
53. Sweere APJ, Luyben KCAM, Kossen NWF (1987) Regime analysis and scale-down: tools to investigate the performance of bioreactors. Enzyme Microb Technol 9(7):386–398
54. Lara AR et al (2006) Transcriptional and metabolic response of recombinant Escherichia coli to spatial dissolved oxygen tension gradients simulated in a scale-down system. Biotechnol Bioeng 93(2):372–385
55. Lorantfy B, Jazini M, Herwig C (2013) Investigation of the physiological response to oxygen limited process conditions of *Pichia pastoris* Mut^{+} strain using a two-compartment scale-down system. J Biosci Bioeng 116(3):371–379
56. Nasution U et al (2006) Generating short-term kinetic responses of primary metabolism of Penicillium chrysogenum through glucose perturbation in the bioscope mini reactor. Metab Eng 8(5):395–405
57. Nienow AW et al (2013) Scale-down studies for assessing the impact of different stress parameters on growth and product quality during animal cell culture. Chem Eng Res Des 91 (11):2265–2274
58. Serrato JA et al (2004) Heterogeneous conditions in dissolved oxygen affect N-glycosylation but not productivity of a monoclonal antibody in hybridoma cultures. Biotechnol Bioeng 88 (2):176–188
59. Papagianni M, Mattey M, Kristiansen B (2003) Design of a tubular loop bioreactor for scale-up and scale-down of fermentation processes. Biotechnol Prog 19(5):1498–1504
60. León-Rodríguez AD, Galindo E, Ramírez OT (2010) Design and characterization of a one-compartment scale-down system for simulating dissolved oxygen tension gradients. J Chem Technol Biotechnol 85(7):950–956
61. Junne S et al (2011) A two-compartment bioreactor system made of commercial parts for bioprocess scale-down studies: impact of oscillations on Bacillus subtilis fed-batch cultivations. Biotechnol J 6(8):1009–1017
62. Sandoval-Basurto EA et al (2005) Culture of Escherichia coli under dissolved oxygen gradients simulated in a two-compartment scale-down system: metabolic response and production of recombinant protein. Biotechnol Bioeng 89(4):453–463
63. Trägårdh C (1988) A hydrodynamic model for the simulation of an aerated agitated fed-batch fermenter. In: BHRA (ed) Proceedings of the Second International Conference on Bioreactor Fluid Dynamics, Cambridge, p 117–131
64. Yu L et al (2012) Hydrodynamic and kinetic study of cellulase production by Trichoderma reesei with pellet morphology. Biotechnol Bioeng 109(7):1755–1768
65. Bannari R et al (2012) A model for cellulase production from Trichoderma reesei in an airlift reactor. Biotechnol Bioeng 109(8):2025–2038
66. Noorman H et al (1993) Measurements and computational fluid dynamics simulation of Saccharomyces cerevisiae production in a 30 m^3 stirred tank reactor. In: Proceedings of International Symposium on Bioreactor Performance, Helsingùr, Denmark

67. Teng ELW, P Kumar, Y Samyudia (2010) Computational fluid dynamics of mixing in aerated bioreactors. In: Proceedings of International Conference on Biology, Environment and Chemistry
68. Elqotbi M et al (2013) CFD modelling of two-phase stirred bioreaction systems by segregated solution of the Euler-Euler model. Comput Chem Eng 48:113–120
69. Morchain J, Gabelle J-C, Cockx A (2014) A coupled population balance model and CFD approach for the simulation of mixing issues in lab-scale and industrial bioreactors. AIChE J 60(1):27–40
70. Lapin A, Klann M, Reuss M (2010) Multi-scale spatio-temporal modeling: lifelines of microorganisms in bioreactors and tracking molecules in cells. In: Wittmann C, Krull R (eds) Biosystems engineering II. Springer, Berlin, pp 23–43
71. Grünberger A et al (2012) A disposable picolitre bioreactor for cultivation and investigation of industrially relevant bacteria on the single cell level. Lab Chip 12:2060–2068
72. Schmidt FR (2005) Optimization and scale up of industrial fermentation processes. Appl Microbiol Biotechnol 68:425–435
73. Humphrey A (1998) Shake flask to fermentor: what have we learned? Biotechnol Prog 14:3–7

Adv Biochem Eng Biotechnol (2016) 152: 153–156
DOI 10.1007/978-3-662-48347-3

Index

A
Acetate, 138
Alanine, 141
Alkylation, 2
Amino acids, 54, 109
  phenylalanine, 91, 109, 138
Ammonia, 36, 68, 76, 82, 111, 112, 138, 141
Antibiotics, 54
Apoptosis, 53
Aspergillus niger, 140, 146
Atomic force microscopy (AFM), 66
Avermectin, 141

B
Bacillus subtilis, 28, 145
Bernoulli's equation, 12
Biohydrogen, 35, 37
Biomanufacturing, in vitro synthetic biosystem, 35
Bioprocess engineering, 137
Bioprocess monitoring, 137
Bioreactors, dynamic environment, heterogeneity, 143

C
Cancer cells, discrimination, PLM, 63
Carbon dioxide, 38, 82, 140, 142
  emission rate (CER), 140
  supercritical, 82
Carbon monoxide, 36
Carotenoids, 138
Cartilage, regeneration, 57
Cell processing center (CPC), 55
Cells, morphology, 58
  processing, automatic, 53
  quality estimation, noninvasive, 53, 57
Cellulose conversion, 75
Cephalosporin C (CPC), 143
Chaotic phenomenon, 142
Clean robot, 56
Clostridium acetobutylicum, 100
Comminution, 77
Computational fluid dynamics (CFD), 24, 83, 85, 139
Cone beam X-ray computed tomography (CBCT), 25
Constraint-based approaches, 91, 99
Corynebacterium glutamicum, 110
Critical speed, 1

D
DAHP synthase, 111
Dark fermentation, 37
Denver flotation cell, 13

E
Electrospray ionization (ESI), 108
Erythromycin, 141, 143
Escherichia coli, 27, 28, 41, 91, 110, 115, 121, 138, 146
Euler–Euler simulation, 139, 146
Euler–Lagrange simulation, 139, 146, 147
Explosion reactors, 81
Extruder reactors, 78

F
Fermentation, 37, 139
  dark, 37
  multi-scale analysis, 139
Fluid dynamics, 137
Flux balance analysis (FBA), 101
Flux variability analysis (FVA), 101, 121
Fluxome, 112
Formate, 138
Fourier transform infrared (FTIR) spectrometer, 70

Fructose, 138
Fructose-1,6-bisphosphate (FBP), 126
Fructose-6-phosphate aldolase, 121

G
Gas flow rate, induced, 15
Gas self-induction, 3
  impellers, 4
  reactor, 1
Gas-induction rate, 1
Gas–liquid contact, 2
Gas–liquid mass transfer coefficient, 18
Geometric flux balance analysis (gFBA), 102, 121, 125
Gluconobacter oxydans, 100
Glucose, 27, 35, 37, 44, 68, 80, 82, 110–121, 138–146
Glucose 6-phosphate, 37, 47
Glutamate, 26, 112, 138
Glutamine, 68, 138
Glycerol, 43, 91, 110, 121–128
Glycerol dehydrogenase, 121
Good manufacturing practice (GMP), 55
Guanosine tetra-(ppGpp), 142

H
Helical ribbon reactors, 83
Hepatocellular carcinoma, 64
High solid loading, 75
Hollow pentagon impeller, 6
Hollow ratchet-type impeller, 5
Hydrogen, biohydrogen, 35
  bioreactors, 35
  detection, continuous, 35
Hydrogenase, 37, 41
Hydrogenation, 2
Hydrostatic pressure, 12, 144

I
Impellers, 1, 143
  Rushton, 113
  self-inducing, 2
  speed, 12, 139
Intracellular fluxes, 98, 120
Isotope dilution mass spectrometry (IDMS), 107

K
Kluyveromyces lactis, 100

L
Lactate, 68, 115, 138
Lignocellulosic biomass, 75

M
Mathematical models, 137, 139, 145
Matrix-assisted laser desorption ionization (MALDI), 108
Mesenchymal stem cells (MSCs), 57
Metabolic control analysis (MCA), 97
Metabolic flux analysis (MFA), 91, 98, 120
Metabolites, extraction, 106
  intracellular, quantification, 124
  quantification, 107
Metabolome, 91, 103
  analytics, 103, 108
  intracellular, 124
  quantification, 91, 107
Metabolomics, 102
  quenching, 105
  sampling, 104
Methanococcus maripaludis, 100
MIA (melanoma inhibitory activity), 69
Microalgae, 27
Microbial kinetics, 137
Mills, comminution, 77
MIR-IRAS, 69
Mixing, 75
Modified double-disk (MDD) turbine, 7
Monod-type kinetic model, 145
Multiple sample preparation, 104

N
Nuclear magnetic resonance (NMR), 98, 109
Nucleic acids, 54

O
Oxygen uptake rate (OUR), 140
Ozonization, 2

P
Pathogen transmission, 55
Penicillin, synthesis, 140, 141
Perturbation, experiments, 91, 95
Phase-shifting laser microscope (PLM), 53, 61
Phenylalanine, 91, 109, 138
Phosphoenolpyruvate (PEP), 109, 126–128, 146
Phosphofructokinase, 127
Phosphotransferase, 146
Pitched turbine downfl ow (PTD) impellers, 7
Pitched-blade turbine (PBTD), 7, 8
Plug-flow reactor (PFR), 145
Polyphosphate glucokinase, 42, 45
Population balance model (PBM), 146
Power consumption, 1, 16
Prephenate dehydrogenase, 121
Pretreatment reactors, 75

Proline, 138
Prostate epithelial cells (PREC), 64
Proton exchange membrane (PEM) fuel cells, 36
Pulse experiments, 95
Pyruvate, 113–115, 121–128, 146

R
Raman spectroscopy, 70
Rapid media transition, 97, 113
Regenerative medicine, 53
Respiratory quotient (RQ), 140
Robots, automated cell culture system, 56
Rocket engine fuel, 36
Rotation speed 11, 00

S
Saccharomyces cerevisiae, 27, 100
Scale-down, 144
Scale-up, 137, 145
Sedoheptulose-7-phosphate (S7P), 126
Self-inducing reactors, 3
  impeller, 3
Shikimate-3-phosphate (S3P), 126, 127
Simultaneous saccharification and ethanol fermentation (SSF), 83
Six-blade impeller, 8
Six-pipe impeller, 7
Soybean oil, 142
Spirulina platensis, 27
Starch, 38
Stator, 11
Steady-state experiments, 91, 96, 112
Steam explosion pretreatment, 81
Steam–methane reforming (SMR), 36
Stimulus–response experiments, 95
Stirred tank reactor (STR), 144
Stoichiometric constraints, 100
Sucrose, 38
Sulfhydrylation, 141
Synthetic pathway biotransformation (SyPaB), 35, 37

T
Thermoenzymes, 47
Time of flight (TOF), 108
Transplantation, automated cell culture system, 56
Tryptophanase, 121
Tylosin, 141
Tyrosine, 109, 111, 121

V
Volumetric mass transfer coefficient, 1

W
Wemco cell, 13

X
Xylose, 38, 46, 80, 82

The manufacturer's authorised representative in the EU is Springer Nature Customer Service Centre GmbH, Europaplatz 3, 69115 Heidelberg, Germany. If you have any concerns regarding our products, please contact ProductSafety@springernature.com

Printed and bound by CPI Group (UK) Ltd, Croydon, CR0 4YY
22/09/2026
02221055-0019